Implementing Quality Management in the Graphic Arts

by

Herschel L. Apfelberg

and

Michael J. Apfelberg

Graphic Arts Technical Foundation
4615 Forbes Avenue
Pittsburgh, Pennsylvania 15213-3796
United States of America
Telephone: 412/621-6941
Fax: 412/621-3049

Library of Congress Catalog Card Number: 95-75372
International Standard Book Number: 0-88362-177-0

Printed in the United States of America

Order No. 1320

A catalog of GATF text and reference books, Learning Modules, audiovisuals, and videotapes may obtained on request from the Graphic Arts Technical Foundation at the address given at the bottom of the title page.

Contents

Foreword

High quality is the goal of every printer. Achieving this goal requires more than just knowing the proper adjustments on the printing press or setting the proper tone reproduction curve in an image-editing software program. It requires more than having equipment that is capable of producing the quality desired. It also requires the human element, e.g., properly trained workers and goal-oriented teams.

In that regard, this book addresses quality more from a humanist viewpoint than a technology-based viewpoint. It doesn't tell you how to adjust your equipment to achieve optimum quality. Rather, it tells you how to manage the process to achieve this end. For example, it stresses the importance of controlling the quality of incoming materials and the importance of working in teams to solve a problem or improve the process.

Implementing Quality Management in the Graphic Arts covers a wide range of topics in its thirteen chapters. The early chapters define total quality management in the graphic arts, describe the tenets of the most prominent individuals in the quality movement through the years, and explain the costs and benefits of a quality improvement program. The book then explains how to collect and analyze data, how to use an SPC problem-solving tool kit, how to determine the capability of a process, how to monitor a process using control charts, and how to design and analyze experiments. The final chapters of the book explain the team approach to continuous process improvement, discuss the implications of change, stress the importance of supplier certification to an overall quality program, describe the Malcolm Baldrige National Quality Award (MBNQA) and how to benchmark a printing company based on the MBNQA criteria, and discuss ISO 9000 and its implications. The authors use printing-related examples to make the various concepts more understandable.

Herschel Apfelberg and Michael Apfelberg, the authors of this book, would like to thank the following technical reviewers for their assistance: Phil Degliomini, J.M. Huber, Inc.; Don Friedel, IonC Associates; Dan Luedke, Keller-Crescent Co.; and Dale Welks, Graphic Industries, Inc.

The authors also would like to thank all the companies and seminar attendees that allowed them to share and shape the information in this book. The book comes not only from the practical experience and research of the authors but also from the experience they gained in sharing this information in professional workshops and consulting projects.

At GATF, the authors would like to thank Thomas Destree, editor in chief, who edited and paginated the book, and David Watterson, graphic designer, who designed the cover.

Frank S. Benevento
Business Manager
GATF Technical Information Group

Dedication

This book is dedicated to Liz Apfelberg, colleague, wife and mother, who exemplifies quality in all her actions.

1 Total Quality Management and the Printing Process

Why Embrace the Tenets of Total Quality Management?

Our technological world is changing rapidly. "Printing," as we know it, is part of this technological world and is continually changing in response to society's needs for communication. The advent of computer systems and our ability to digitize and manipulate information are examples of the changes taking place. As these evolutions and revolutions continue, we need a work force that can adapt without being threatened. The work force will need to be trained with the skills necessary to deal with change. These skills will become as important as, if not more important than, the skills needed to complete the actual printing job. In other words, let us remove the barriers to adopting change by educating and training the work force to participate and share in the inevitable technological changes that will continue to take place.

The skills required for a changing work force are different than the skills required for status quo environments. Traditionally, printing required employees who had the ability to learn a skill and continually improve that skill. A typesetter became better and better at perfecting the skills of speed and accuracy, while the press operator became better at accurate makereadies and run performance. New technologies require the ability to develop systems quickly, anticipate customer needs, and translate them into workable specifications, while at the same time trying to anticipate what must be changed in order to compete effectively. These new requirements will necessitate a work force that is team oriented, continuous process improvement centered while at the same time anticipating, planning, and sharing in the re-engineering of the system to better satisfy the customer. This is part of the process of meeting new customer needs, not merely satisfying old ones.

Those printing organizations that grasp the basic tenets of total quality management (TQM) and invest in the time and effort of re-engineering their organizations to embrace this philosophy will be the leaders of the technological revolution. Employees will then not feel threatened by these rapid changes as their skills will be centered around participating in these changes for customer satisfaction. They will know that initiating as well as accepting change will mean that their jobs will not be threatened and that their contributions will not only be welcome, but absolutely necessary. This is the basic reason to embrace the principles of TQM.

However, beginning the journey to a total quality management company is hard work. It requires a plan that is long term while having short-term goals. Do not expect employees to leap upon the TQM bandwagon. When first introduced, many people get excited about the prospects that TQM offers. But coaching rather than directing is a large change for managers. Learning how to take responsibility for quality, waste reduction, and process improvement can be very threatening to the work force. Successes and failures belong to everyone in a TQM organization.

It takes top management commitment, long-term objectives, short-term goals, and perseverance to make TQM work. Once committed, top management must hold everyone to the TQM process. There will be setbacks, but only top management can stay the course. Top management must understand that TQM is a re-engineering concept that requires the use of continuous process improvement. Most experts agree that it takes at least three to five years to make the transition from the standard operating procedures of most printing companies to the acceptance and implementation of a TQM organization. With this in mind, planning and patience are required to be successful. TQM is not a program or quick fix for your immediate problems. What TQM will do for your company is to give it a long-term structure for dealing with the critical areas of re-engineering, continuous process improvement, benchmarking, teamwork, empowerment, the use of data for decision making, and a central focus on the ever-changing needs of internal and external customers.

Total Quality Commitment from Top Management

It is extremely difficult, if not impossible, to implement TQM without commitment from the leaders of a printing organization. Top management needs to take the time and effort to understand what a commitment to TQM means before introducing it to their organization. They must understand the tenets that comprise a philosophy of TQM and how they will have to implement them in their organization. TQM is a way of life, not a program. It takes time to convince and then train everyone in the organization to the tools necessary to sustain a TQM process. Once committed to the philosophy of TQM, it takes time (which translates into money) to not only introduce TQM elements but to do the much harder task of sustaining the effort. Top management needs to take the time to research the benefits of committing their company to the TQM process. It is hard work and requires an understanding of what TQM is and how it relates to the company. A suggested plan for determining how TQM can be implemented is to first study the philosophy of TQM; second, develop a plan of how it can be applied to your company; third, determine the financial investment necessary to initiate and sustain the TQM effort; fourth, assess the financial investment the organization has to invest to initiate the TQM effort; fifth, determine how long it will take to start to see the payback from the TQM effort and what that payback should be. When these have been determined top management needs to develop a plan, follow it through, make adjustments and continue on the TQM road. Only through top management commitment can a printing company become a TQM-driven organization.

Suggested TQM Plan

- Study and understand philosophy
- Develop a personalized approach
- Determine financial investment vs. capital available
- Determine length of time for payback

What Is TQM?

All or Nothing Approach to TQM

Although this book offers a fairly complete approach to implementing total quality management, it is suggested that printing companies embarking on or exploring quality improvement processes evaluate their particular needs and which particular TQM element or elements will allow them the greatest return on investment both in the short and long run. This approach will assist in energizing the organization in a belief in TQM strategies and also assist in convincing or reinforcing the financial value of this process.

The suggested approach is to develop a strategy that will allow the organization to grow and prosper. To embark on all nine items discussed below can overwhelm and lead to early disappointment.

Basic Elements of TQM

The elements that comprise TQM include focus on the customer, involvement of the entire printing organization in the process, an employee team effort, a mind-set for process improvement, empowerment of employees to get the job done, benchmarking for process improvement, development of partnerships with internal and external suppliers and customers, re-engineering the organization and its processes when necessary, and the measurement and management of quality.

Total Quality Management Concepts

- Focus on ***customer*** (internal and external)
- Involvement of the ***entire*** printing organization
- Requires a ***team effort***
- Empowerment of ***employees*** of the printing company
- Develop a mind-set for ***process improvement***
- ***Benchmark*** for process improvement
- ***Partner*** with suppliers and customers
- ***Re-engineer*** where needed
- ***Measure quality*** so that it can be managed

The Customer, Both Internal and External

Total quality management is driven by the realization that your business success is dependent upon serving your customers better than anyone else can. The customer is usually viewed as the one who purchases the printed product or, in the case of the final customer, the one who uses the finished product produced by the printing company—e.g, brochure, book, magazine, label, package, etc. These customers are referred to as external customers. However, there are customers who are part of the process that goes into making the final product or subcomponents of the finished product—e.g., sales request, estimate, artwork, assembled image, color separation, printing plate, printing, and finished product or the delivery of this product. These customers are referred to as internal customers. Everyone in the loop from initial order to delivered product is a supplier or a customer to someone else in the process. This concept is critical to TQM as it empowers each and every individual involved in the process to make sure that they totally understand the needs of their customer. These needs must be translated into specifications that can be met or exceeded on a consistent basis. Any slippage in the steps can and usually does cause an inferior product to be manufactured.

TQM Is Not Just for the Printing Production Process

TQM encompasses all the elements that go into making up a printing organization. It would be a horrendous mistake to undertake a TQM philosophy in your printing organization without understanding the total relationship between departments from marketing and sales, art and design, computerized and conventional prepress to accounting, estimating, production planning, the manufacturing process, and delivery of the printed product.

The reason that this would be a horrendous mistake is that each person and department in the company must understand their personal importance to customer satisfaction. When a job is rejected or the customer requests a price reduction or even worse if the customer goes to your competitor for new or remake work, it is never because the job is defective in every way possible but because some specific defect (or defects) is not acceptable to the customer. Any one department can do a bad or good job and it does not matter; it is the entire job that is accepted or rejected. TQM is putting together all the small elements so that the entire job is excellent and the customer is satisfied.

Team Effort

Because the final product usually requires more than one person to complete the job, it is important that a team effort be fostered. It must be understood that team efforts have not been part of the American work culture. The reward system that most of us grew up with rewards individuals for their particular effort regardless of what the group does. It is difficult to initiate teams into any company, however it is essential to initiate and sustain the team effort as this will allow for the development and maintenance of internal vendor and customer specifications. Other benefits of team efforts can include process improvement procedures, increased enthusiasm to implement what one has been part of developing, and viewing the printing job as a total product and not just small elements.

There are different types of teams for different purposes. The executive steering committee is comprised of the top leadership of a company. This committee is an ongoing team that oversees the entire company's TQM effort. Another type of team is a functional team that is comprised of people within a given department and has the ongoing responsibility of initiating changes within their particular area. The third type of team is the cross-functional team that is comprised of people from various areas within the company and can on occasion have members from outside suppliers and customers. They have the responsibility to resolve problems that are situational and are generally given problems from the executive committee to resolve. Cross-functional teams are generally discharged at the end of a particular problem resolution. The fourth type of team is the self-directed work team made up of a group of individuals who ordinarily work together on a daily basis to get a job done. The work is designed to give the team control of the situation. As a self-directed team, they are responsible for various management and leadership roles: they plan, control, and improve their own work processes, they set goals and inspect their own work, they set schedules and review their own performance, and they set their own budgets and coordinate their work with other departments. They are responsible for ordering materials, keeping inventories, and dealing with suppliers, and they are charged with their training and updating needs. Self-directed teams also evaluate individual and group progress toward an agreed-upon goal. Self-directed teams require supervisors who are coaches and facilitators.

Teams must be trained in how to function, and they must be given the tools needed to work together effectively. Members must be able to offer suggestions and resolve particular problems in a timely and cost-effective manner. For success in team usage it is imperative that top management plans how a company will use teams and determines the outcomes that are desired. Teams can not be set up without being given the proper time to train, time allocations for results, and financial support to sustain their efforts. It is important to have clearly defined missions and time frames for feedback from the teams.

Mind-set for Process Improvement

Traditionally, the printing industry has not taken continuous process improvement seriously. As printing has been a highly skilled craft, it was generally assumed that the printing process would remain fairly constant. Printers in general were into continuous content improvement; i.e., getting better and better at doing the same thing. Process improvement implies always looking for a better way of doing things.

Some of the objectives of process improvement are to build quality into the product by looking for more consistent procedures, reducing inspection wherever possible, shortening the throughput time of printing jobs by analyzing those steps that are not value-added, and to be cost-effective by reducing waste and other nonadditive components of the printing process.

Process improvement requires a mind-set for change. It requires that management and employees not only be open but also continually question each step in the printing process, from the sale of the job to its delivery including its manufacture, billing, and customer payment. Process improvement requires that you understand each step in the total operation of your organization and how these steps are interconnected. Detailing the steps in the process should be done by the people doing the actual work from sales personnel to the janitor. This can be done through the use of flow diagrams (also known as process maps). Processes involving more than one person should be done by the entire group involved in that particular process. It is necessary to first determine what is being done presently before working on process improvement steps. The vital questions are how to improve throughput and quality while reducing cost and satisfying customers.

Empowering Employees

The empowerment process not only allows employees the opportunity to participate in how their work areas operate but also necessitates that employees take ownership of their individual work areas and develop an ownership outlook toward their jobs. The empowerment process requires that managers take on a coaching and facilitating role and become aware of how they can assist employees with their particular needs to get the job done better, faster, and easier. The empowerment goal is to improve quality, productivity, and service. To empower the work force means more than listening to requests for changes in materials, working procedures, equipment, training, communication systems, and other areas that will assist the employee in doing the job more effectively. Empowerment means encouraging and training the work force to take responsibility for making decisions relating to quality, productivity, and service to the customer.

Benchmark for Process Improvement

Benchmarking is the process of measuring your company's performance in each function and comparing that level of performance with the level of performance achieved by successful leaders. There are three types of benchmarking:

- *Internal benchmarking,* which occurs when someone or a group is performing in a world-class manner that can be used by the entire organization as a benchmark for process improvement.
- *Competitive benchmarking,* which occurs when a printing company is doing similar work and is worthy of being emulated in one or more of the areas of how it conducts its business.
- *Generic (world-class) benchmarking,* which occurs when any organization that is not in the same field has attributes that can be emulated by your organization in order to become world class.

Benchmarking can and should be in all areas of your organization including, but not limited to, the primary areas of manufacture as well as the areas of marketing, sales, billing, training, maintenance, and inventory control.

Organizations that specialize in collecting and disseminating data that can be used for benchmarking purposes include the Printing Industries of America (PIA) and the National Association of Printers and Lithographers (NAPL). Both organizations do comparative national and regional studies in various areas of printing management. These studies can be used as an initial benchmark as to where your organization stands in relationship to these regional and national statistics. The Graphic Arts Technical Foundation (GATF) and other technical organizations offer studies and published materials on best printing practices, which can also be used as benchmarks. However, one should also study successful local businesses for tips on how to improve your organization. Local banks, supermarkets, and other local businesses sometimes have world-class operations that can be used to model certain aspects that can improve your company's competitiveness. Organizations that are not in direct competition to your company are generally more open to sharing ideas on how they achieved a specific world-class attribute. Make it a point to always study how others do business so that you can compare your company's practices against theirs and continually work towards process improvement.

Partnering with Suppliers and Customers

Partnering means working together for the common good. When companies purchase their printing materials and supplies by price alone, the ultimate cost may far and away exceed choosing suppliers for their service, quality, and price. When you have to wait for materials that are needed immediately or when they do arrive and are in poor condition, the hidden costs are excessive. Supplier partners are those who are always looking out for your organization, assisting you to become better by offering materials, supplies, and services that can be depended upon. These are the suppliers that will suggest new and different materials and procedures and will offer an overall more intimate working relationship. It is a win-win relationship that needs to be nurtured. The supplier can depend upon your business, and in return you can be assured that your needs are being met by experts.

What is true of supplier partnerships is also true of customer partnerships. To maintain customer partnerships requires that you have an intimate knowledge of the business needs of your customer. You need to be constantly vigilant in keeping abreast of your customer's present and future needs. The questions you need to be constantly asking are "What are your customers' competitors using by the way of printed matter?" and "What new strategies in print media will make your customers more competitive?" This can be a win-win environment. Your customers do not have to continually look for new printers, and you have steady customers.

In the long run, partnerships allow more time to explore other value-added activities. Rather than looking for new suppliers and customers, you can benchmark attributes to be more productive. Spending time on activities of process improvement and not on the hunt for new suppliers and customers are the rewards of supplier and customer partnerships.

Re-engineering the Printing Organization

Re-engineering is radical change, while process improvement is incremental in nature. Finding faster or more user friendly prepress films is process improvement. The moves from darkroom compositing to a daylight environment or from lith chemistry to rapid-access chemistry for film development were process improvements. However, when we bypass the entire darkroom process and go from design concept, generated in digital form on a computer and scanner, directly to press with this digital information, that is re-engineering. Obviously, these advances make our jobs easier and the end results more dependable and economical.

Re-engineering is not throwing out the baby with the bath water. Re-engineering changes the basic structure of the system in question. Re-engineering in concept is starting with the same baby but fresh water. If the printing industry is part of the communication field, then the big question is, "How do we communicate in the most inexpensive, quality fashion and remain customer-driven?" Why even think of having a hard copy, why not use E-mail or CD-ROM? The answers will be mandated by total customer focus.

A printing company's re-engineering focus needs to be based on two related factors: total customer satisfaction and effective and efficient internal processes. We must be continually vigilant in the pursuit of assisting our customers to communicate better. That is our job. The dilemma remains, "How do we remain specialized and niche-focused while continually looking to satisfy our customers' needs?" However, offering products and services that will no longer be needed will not keep you in business.

Internal re-engineering is not only a technical emphasis, but also implies restructuring how the company operates. In the past, close supervision and craft skills kept the printing industry competitive. Today, there is a need for people to relate more to the final product, and teamwork is essential to that cause. Close supervision does not work with teams, which function best with coaches and facilitators. Therefore, the printing company of the future will flatten its organizational structure by having supervisors take on the role of coaching and facilitating rather than directing and inspecting.

Measuring Quality So That It Can Be Managed

There are many reasons for applying statistical process control (SPC) to the printing TQM process. The following seven purposes of statistical process control describe the critical reasons for the use of SPC in the printing organization:

- **To build quality into the process by preventing defects rather than detecting defects after they have occurred.** Building quality into the printing process necessitates an understanding of what the customer perceives as quality. In many cases, the printer needs to educate the customer as to the capability of the printing process in relationship to the end product needs. This

Purposes of Statistical Process Control

- ***Prevent*** rather than detect defects
- Indicate the need for ***corrective action***
- ***Continuous*** monitoring of the printing process
- Direction for ***process improvement***
- ***Quantitative proof*** of quality
- Identify the sources of ***variation***
- Determine ***process capability***

perception must then be translated into specifications that are achievable. The specifications must include incoming materials as well as the printer's production specifications. It is imperative that a printing company's suppliers understand what is required by the print buyer. This not only includes ink and paper suppliers, but also artists and designers of the printed product. Working closely with buy-out suppliers is also critical. Depending upon whether you are a full-service printer or a niche specialist, the buy-out suppliers must also know the specifications of the print buyer. Separating a job for a newsprint application when it will be printed on coated paper will only lead to a defective product. It does the printer and the customer no good for the price to be reduced or the job rerun because an incomplete or inaccurate specification was followed. The objective is to determine all the specifics of a printing job from the customer and translate these qualitative needs into quantifiable information that can be followed by print designers, suppliers, and printers. When using statistical process control procedures, the end result is preplanned. Sources of variation that affect the printed product must be analyzed, and the necessary steps taken to reduce and eliminate these variations. It is imperative that one understands the inherent variation in the process before determining specifications. The process is then monitored for conformance to these specifications.

- **To indicate the need for corrective action and enable us to make sound decisions about what action to take.** A plan that allows us to collect data, chart that data, and analyze the information gives the statistical process control user an organized method for taking corrective action effectively. The use of statistical process control enables us to separate feelings from facts. We are not suggesting that feelings can not be rooted in facts. However, unverified feelings can lead to costly changes that might not give an organization gains in cost-effectiveness, productivity, service, or quality. A printing company may not even be aware that a change is taking place. Sometimes variables can change in small, undetectable increments if not monitored, evaluated, and corrected. Monitoring a printing press's rubber roller durometer over time will assist in detecting when a rubber roller will get hard enough to cause dot gain or slur in the finished print. Changing out-of-specification rubber rollers will eliminate the durometer attribute as it pertains to dot gain and slur. Monitoring incoming materials for specific variables that are important to the process will assist in determining how consistent suppliers are and if their materials enable you to continue to produce quality products. Being able to accurately assess the cause of a problem in a timely manner can save the user of statistical process control procedures from producing defective products.

- **To monitor and evaluate the performance of the process continuously in order to know if it is deteriorating or improving.** After the printing process is determined to be statistically stable, specifications are developed and agreed upon. To maintain this stable state, it then becomes important to continue to use the tools of statistical process control. The monitoring system uses the data that is developed as the process is evolving. All aspects of the printing process need to be monitored. Customer needs and expectations are continually being upgraded, requiring customer surveys—both formal and informal—and an astute sales force to keep abreast and communicate these needs to the company. In this manner, products and services evolve that keep customers satisfied. The quality of paper, ink, plates, and film can change as a function of the supplier's process. Printing equipment is susceptible to wear and needs to be continually monitored for performance and the need for preventive maintenance. One can not let up on monitoring and evaluating the entire printing process. The system to be set up must be practical and usable and needs to include continual reduction of variation, what is to be monitored, what the specification is, who will do the monitoring, how often will the monitoring be done, and what will be done with the information after it is collected and reported.

- **To provide directions and ideas for the improvement of the process.** The use of the tools of statistical process control allows for analysis of trends and logical adjustments in the process. Charting how many calls the sales force makes to potential customers before getting a particular order allows us to analyze the effectiveness of our sales strategy. Keeping and using data on customer satisfaction—be it cost, service, quality, or any subcategory of these major customer satisfaction determinants—can reduce turnover of the customer base. A proactive stance can be taken toward customer needs, thereby reducing the risk of loosing customers. The outcome of such a practice is to allow for the development of more long-lasting relationships with customers. This gives each party the opportunity to work on continuous specification and communication improvements, which inevitably lead to more consistent quality, service, and cost containment, or preferably cost reduction. This also allows for growth by attracting new customers while maintaining present ones.

- **To provide evidence and quantitative proof of quality.** To claim that an organization is a quality manufacturer of printing without having evidence to support this statement can cause difficulty in maintaining and getting new customers. Customers are becoming more sophisticated in what quality means to them and are asking for proof that their print suppliers can do what they say they can. How does a printing company conduct itself in a quality fashion without tracking its processes through data collection and analysis? It is difficult, if not impossible, to know what is happening without monitoring the entire process. How can a potential or existing customer be shown that your process provides for predictable and consistent results without supportive evidence? Statistical process control is a scientific method giving the printer a practical management tool for the orderly collection of data that is supportive of the quality mission.

- **To identify the sources of variation and the causes for defective goods and services.** By collecting data of the printing process, using the scientific methods of statistical process control, and analyzing this information, one would be able to determine the variation inherent in the process. Isolating the causes of variation can be extremely difficult if one does not keep records of what has taken place. The toner cartridge on a copying machine has a finite quantity of toner. If high-speed copying is being done and the cartridge does not have enough toner in it to finish the job, the job may have to be rerun. However, by using process capability procedures of quantifying the life of a cartridge and by using a counter to assist in determining when to change the cartridge, the problem will be eliminated. Identifying and reducing the source of defective goods and services is one of the most important benefits of statistical process control. It is a means of improving the process, which will lead to consistent quality, cost reductions, and satisfied customers.

- **To determine process capability.** Most printers do not know the true capability of their printing process. In the development of specifications, it is important to know what each piece of equipment can do in terms of the important attributes as determined by the customer. If accurate folding is determined to be the most important specification of a brochure, it is essential to determine what is meant by "accurate" in quantifiable terms. Is the print-to-fold accuracy $\pm\frac{1}{16}$, $\pm\frac{1}{32}$, or $\pm\frac{1}{64}$ of an inch? Once this is quantified, the machine needs to be tested to determine if it is capable of delivering this specification. Without doing a capability study, which requires data collection and analysis, it is hard, if not impossible, to determine the capability of the folder. Process capability is dependent upon equipment design, maintenance, people skills, materials, methods of operation, and the environment at a particular moment of time. Understanding process capability will assist in determining the type of work that the organization should be doing in relationship to its specifications. Profitability of printing jobs depends in part on the organization's ability to do work efficiently and consistently. Quantifying all aspects of the printing process through capability studies can assist in determining whether the right equipment, materials, people, and total process is inadequate, adequate, or more than adequate in relationship to product specifications. Using the tools of SPC to identify and assist in reducing variation in the process will also lead to increased productivity and profitability. The organization's capability of fulfilling the needs of the present customers while assisting in determining capabilities to enter other printing markets can be analyzed with the use of statistical capability procedures.

Introduction to SPC Tools

The seven distinct statistical process control tools should be visualized as building blocks that are related and comprise an additive system. These tools have specific applications and assist in analyzing and managing the operational processes of the printing environment. The printing environment includes all aspects of a company from top management activities to marketing, sales, print manufacturing, and distribution. It is suggested that an organized systematic approach be used in understanding how each of these tools can be used to improve the printing organization. Many organizations have taken the approach that they could start with any of the seven tools and be successful. This is not true, and it is suggested that a company be extremely careful in approaching the use of advanced tools until totally familiar with the use of the basic

The Use of Statistical Process Control in a Total Quality Management Printing Company

Facts and Data

▼

Basic SPC Tools
Flow Charts
Cause-and-Effect Analysis
Check Sheets
Pareto Analysis
Histograms
Run and Control Charts
Correlation Diagrams

▼

Management by Facts and Data

▼

Continuous Organizational Improvement

tools. It is more comfortable and easier for everyone to accept statistical process control if we start with flow charts and work toward the use of correlation diagrams and control charts. It will require patience, but the results will be long lasting.

Flow diagrams or **charts** (also known as **process maps)** are used to assist in systematically breaking down the organizational process into a step-by-step picture of each component. Symbols are used to indicate activities, decisions, beginning and ending points, and the flow that the process takes. Flow diagrams can be used for each element in a printing company. Examples of these uses may include how a customer service representative interacts with customers, sales, and manufacturing as well as how a color scanner is utilized to separate transparencies into printing negatives.

Cause-and-effect diagrams (also called **fishbone analysis** and **Ishikawa diagrams)** are used to break down a process into the major categories that affect the process. These include the environment, people, materials, methods, measurement, and machines. These six areas define all the possible causes of process variation. Cause-and-effect analysis is a natural step between flow charting and the actual gathering of data. This tool can be used in all areas of an organization, not just the manufacturing cycle. Applications range from identifying quality to ensuring well-trained and well-educated employees.

Check sheets are tools that allow you to graphically collect data. When a well-designed check sheet is used, it allows one to collect, compile, and analyze information. Five types of check sheets are the production process distribution check sheet, the defective item check sheet, the defect location check sheet, the defect cause check sheet, and the confirmation/inspection checklist. Each of these check sheets can be modified and personalized for any part of the printing process.

Pareto analysis is developed around the basic concept that 80% of a specific effect is due to 20% of the causes (80–20 rule). It allows you to separate the vital few "real causes" of a problem from the trivial many. A good use of check sheets is collecting data from a specific problem area. A bar graph is then developed for easy visual analysis. Areas in which Pareto analysis have been used with great success include inappropriate use of a sales force, inaccurate estimates, customer dissatisfaction, causes of paper waste, and job reruns.

Histograms are bar graphs that are used to illustrate the distribution of data obtained from a process. It is a pictorial representation of the variation of a set of data. Each bar represents a range of data with the height of the bar determined by the number or frequency of individual readings in the range. Histograms are important in analyzing the distribution in terms of its shape, spread, and centering. The histogram is an excellent tool to use to determine if the process is capable of meeting specifications. When the data is normally distributed, the histogram will resemble a bell curve. Histograms can be beneficial in providing explanations of the printing process. However, they also have their limitations, which include the fact that they do not take time into consideration and do not indicate trends.

Run charts visually display the variations in a process over time. The run chart displays an average line taken from the data and will have data points and lines connecting them. In this manner the user can readily see how each data point varies from the average. Run charts can also be used in most of a printing firm's departments to plot events over time. Some of these areas include sales calls per month, completed jobs per week, estimates per week, ink density per job, and customer complaints per year.

Control charts are run charts with statistically calculated upper and lower control limits. The control chart pictorially shows ranges and averages over a given time. It can be used to monitor, analyze, and control either an attribute or a variable. The control chart can distinguish between special and common causes. It can reveal when a process has changed. It describes what is going on within a process graphically. Control charts are very useful in monitoring and assisting in the control of printing processes that are capable and have specifications. The control chart has statistically calculated lines to illustrate the process upper and lower boundaries of anticipated variation. Some uses of the control chart are in monitoring printing press and bindery dynamic running operations.

Correlation diagrams are used to determine the strength of the relationship between two variables. The use of correlation diagrams usually comes from the information that you get from cause-and-effect and Pareto analysis. The correlation diagram will verify the strength of the relationship between two variables and assist in predicting modifications to the process. It allows the printing organization to optimize its operating procedures. Some areas of use include training dollars vs. productivity, turnover rate vs. number of employees involved, sales vs. on-time deliveries, and ink density vs. dot gain.

Expanded Definition of Quality

Quality is not confined to producing a quality product. Instead, quality must be defined as encompassing all aspects of *"full customer satisfaction."* This expanded definition of quality includes three basic views. Each view and its relation to quality is ultimately determined by the needs and requirements of the customer.

The following chart identifies three perspectives that when applied to a printed product make up those quality assessments that identify *"full customer satisfaction."*

Overview of the Expanded Definition of Quality

Perspective	Key Factors
Product-based view "Grade" or "inherent value"	Durability, reliability, maintainability, cost, price, accessibility of service, or some other product feature that affects customer satisfaction.
User-based view "Fitness for use"	Performance requirements, safety life cycle, delivery and quantity, operable as specified, competitiveness, longevity.
Maker-based view "Conformance to specifications"	Design specifications, conformance to requirements, producibility, inspectability, employee morale and safety.

The **product-based view** determines whether the end product meets the perceived needs of the customer. A printer may deliver an end product that is totally satisfactory to the customer, however the delivery is two weeks late. If the customer is time-sensitive, and most are, and is working with a just-in-time (JIT) inventory system, the delivery time of this product is well out of specification, and the quality is not within the acceptable range of your customer's quality expectations.

The **fitness-for-use** characteristic requires that the end user receives a product that is usable as specified. The product that is produced may fall well within the limits of customer satisfaction, but what happens if the delivery system is destructive to the product, if rolls of paper are torn by grab trucks or boxes, or the finished print product is damaged?

As a printer or supplier, you must make sure that you understand the specifications that are required to make the product a useful end product. For example, the color separator who uses predetermined dot gain curves and ink hue values that have been developed for SWOP (Specifications for Web Offset Publications) and does not understand how to modify these values for a specific end product need can not meet the needs of the printer and will not **conform to specifications.**

Quality is a combination of the perspectives of the **product-based view** that deals with the inherent value of the product, the **user-based view** that concerns itself with the fitness-for-use aspects of the product, and the **maker-based view** that deals with the specifications that go into a printed product that ultimately affect the end user. When these three perspectives of quality are fully developed and dealt with, the result is a quality product and a satisfied customer.

Organizational Activities for Quality

The management of a printing organization must first understand what determines full customer satisfaction and from this frame of reference lead the organization in understanding the quality mission. Management must lead the effort in determining what quality is, quantify it, and maintain a full customer satisfaction focus. The following eleven steps identify the organizational activities for quality.

1. *Full customer satisfaction.* The customer defines what quality means. This becomes the driving principle of the printing business.
2. *Quality is first and foremost a top management issue.* Management must believe that quality means "full customer satisfaction" and that quality is the cornerstone of a successful printing business.
3. *Quality first policy.* Management must sustain an attitude and practice a "quality first attitude." The "it is good enough this time" approach to doing business must be completely irradicated. TQM is an evolution and must be worked at by making quality strides and improving upon the process in a consistent manner.

4. *Management must:*
   - Understand the full scope of their quality mission.
   - Develop policies and encourage procedures that will ensure quality.
   - Set annual, if not more frequent, quality improvement goals.
   - Establish methods to achieve these goals.
   - Develop and lead the quality improvement process.
   - Develop cost-of-quality data and its evaluation.
   - Develop an atmosphere free of fear of failure.
   - Involve everyone in continuous quality improvement strategies.
   - Allocate financial support for initiating and maintaining statistical process control as part of the total quality control process.
5. *Training must be ongoing.* Management must ensure that all employees are educated in the need for quality and given the appropriate tools to accomplish the requirements of their jobs.
6. *Collect and use data for analysis.* Use statistical process control. Use data, not emotions, for problem solving.
7. *Build quality into the process.* Work toward building quality into the process and not inspecting it out. Develop systems that assure quality at each step of the printing process. Specific standards and their measurement are required at each stage of the process. Control the inputs to assist in outputting a quality printed product.
8. *When a defect occurs.* Remedy it first and then seek out the cause and take action to ensure that the defect never occurs again.
9. *Do it right the first time, and do not pass on any nonconforming goods.* Use the principle of internal and external customer specifications so that the "next person and/or process is your customer."
10. *Total quality management involves everyone.* The entire organization is responsible for working toward reducing variation, making the entire process predictable and thereby ensuring that 100% of the goods and services passed on to internal and external customers conform to specifications. Quality depends on the coordinated efforts of the marketing, sales, design, engineering/maintenance, purchasing, manufacturing, quality assurance, packaging, shipping, and service departments.

11. *Continuously review the quality mission and its process.* Quality is an ongoing process and needs to be reviewed for changes in customer needs and whether the process is in control.

Summary

In this chapter, we have seen that the main characteristic of those printing companies embracing TQM is that they are focused upon the present and future needs of their customers. This includes not only the final customer of the printed product but also intermediate customers (agencies), internal customers, managers, supervisors, and workers. These companies expend a great deal of effort in listening to the needs of these groups and determining procedures that can fulfill these needs. The organization as a whole fosters coaching, mentoring, and facilitating at every level. Everyone involved with the company is looked upon as a team member. These members include suppliers, company employees, and customers. Whenever possible the company nurtures partnership relationships where all members focus on process improvement using the tools of total quality management.

2 Foundations of the Modern Quality Movement

Manufacturing Categories

The printing, publishing, packaging, and allied fields are very diverse in their manufacturing techniques. Because of the various components of this industry, it is important to understand the differences in how manufacturing is accomplished. By doing so one gets a fuller understanding of the historical development of total quality management and its relationship to this industry.

Manufacturing can be broken down into different procedures and flow. These categories are the continuous or flow, assembly, job shop or custom, and project-oriented systems of manufacture. Many printed products and their subcomponents can be classified as containing various components of these categories. It is therefore difficult to make generalizations that printing falls into a specific manufacturing category.

Printing Manufacturing Categories

- Project—one-of-a-kind jobs
- Job shop or custom—various machines needed to complete the product
- Assembly—one product, assembling various pieces
- Continuous or flow—complete product from one machine

Project-oriented process. Graphic arts lends itself to one-of-a-kind work. Traditionally, the building of a bridge or an aircraft carrier was considered project-type manufacturing. Each project had unique differences and requirements that demanded specific individual attention. The personalization that some segments of the graphic arts require, in part, necessitate a project orientation. Artwork and manuscripts are generally considered individual projects, one-of-a-kind work. With the advent of the personal computer and desktop publishing, project operations are becoming integrated into job shops, assembly, and ultimately printing manufacturing methods.

Job shop or custom process. These operations tend to have pieces of equipment that are capable of customizing products for the individual needs of a client. Commercial printers tend to be job-shop-oriented. A commercial printer traditionally had the capability of supplying business cards, letterheads, business forms, brochures, and any other printing need of the customer. The commercial printer is an expert at being a generalist. Employees have to be very versatile, if not expert, at most aspects of printing. Equipment is not necessarily kept busy, but the commercial printer has to have a wide variety of equipment in order to meet the many needs of the customer. Job shop operations in printing have been scrutinized in the last several years, with most people agreeing that printers should find niches in which they can excel. This will move the job shop orientation toward the assembly type of operation.

The assembly process. The assembly process is used when one or more items come together to make a completely different product. Car manufacturing is a classic example of an assembly type of operation. Assembly-line operations generally bring together products that have completed subcomponents (printed and folded signatures or printed covers). Assembly type of manufacture is thought of as being connected to an assembly line where the product flows at a prescribed speed while workers affix parts, tighten, or otherwise add to the product as it proceeds to its final destination. The equipment that puts signatures one on top of the other and then adds a cover, saddle-stitches these components together, and finally three-knife-trims them is an example of the assembly process.

The assembly process is generally labor-intensive, requiring coordination of just-in-time subcomponent planning. However, efforts are always being made to automate the assembly process, making it more like the continuous process. The use of digital information, from the desktop publisher, to control the operations of the printing press operations and integrating press operations with bindery processes continually erode the use of the assembly process. Some directory and magazine production processes are examples of the assembly process.

Workers have traditionally been isolated from one another in what they do and how they do their work. Sometimes, it is hard for individuals to visualize how their labor impacts on the completed product. Assembly operations in the graphic arts are growing as we move away from hand operations and job-shop-orientation printing.

Continuous or flow process. Continuous-flow manufacturing assumes that the product is made in one continuous operation. Papermaking is a continuous operation where the cellulose fibers, as a slurry, flow out onto a fourdrinier wire and end up at the other end of the machine as dried, cohesive paper. Web printing presses come close to simulating this means of production. Paper enters the machine and can be converted into a printed product that can be folded, glued, cut, perforated, embossed, and assembled in one operation. The ultimate example of this process in printing is the belt press where a roll of paper is put on one end of this machine and a finished book is delivered at the other end, in one operation. The printing of newspapers, advertising inserts, and business forms also fall into the category of continuous manufacturing.

Flow operations require that trained individuals work together in an organized, cohesive group to control the continuous process. The operation of a large web press may take up to nine individuals, all of whom must be coordinated into one cohesive team in order to print a quality product at speeds up to 70,000 signatures or finished products per hour. Continuous processes are naturally team-focused because a group of specialized people who coordinate their efforts is required to properly and effectively operate the machine.

Summary. The trend in graphic arts has been away from the generalist job-shop orientation and toward companies that develop niches. There has been an integration of the assembly and continuous manufacturing techniques with the advent of more automated types of assembly equipment. Henry Ford developed the assembly line as a means of breaking down complex tasks into simpler procedures, thereby minimizing the skills required to assemble the car. The printing industry is continually re-engineering its processes to smooth out the flow of operations and decrease the steps required to produce a printed product.

Today, the graphic arts industry is tackling these changes through technological developments that allow the consolidation of highly skilled crafts. The integration of typesetting and prepress and the integration of color separation and image assembly are examples of combining two highly skilled areas into one skilled craft.

The development of finishing operations at the end of the web printing press has made this operation look more like continuous manufacturing. Book printing presses that take a web of paper and convert it into a printed, bound, and delivered finished product also compare favorably with continuous manufacturing. The advent of the desktop computer has integrated the printing process with the graphic designer and the prepress operation, which may eventually eliminate the need for prepress as we now know it.

The future will call for us to analyze print manufacture as an integration of continuous, assembly, job shop, and project-oriented types of operations. Total quality management and its concept of integrating the entire operation into one systematic related system will become the imperative. A team orientation is required if the designer is to control the preparation of information that will be taken directly to the print manufacturing facility with little or no further manipulation by the printer. We must understand the strengths and weaknesses of the continuous, assembly, job shop, and custom manufacturing processes to better use the tools that total quality management has to offer in maintaining and improving the graphic arts company. It is also important to study the contributions of the leaders in the total quality management movement and how their theories impact upon the printing industry.

Some Dominant Quality Thinkers

The following individuals have contributed in both theory and practice to statistical process control and total quality management. Today, many people give credit to one or maybe two individuals and follow their teachings religiously. We believe that a better understanding of, and ability to implement, total quality management techniques will occur if one takes a holistic view of these theoreticians and practitioners. Some common threads that pervade their thinking include the need for the collection and analysis of data, a belief in the scientific method for analyzing problems, and, in the final analysis, a belief that it is people that make business happen. In the long run, profitability in the printing industry is dependent upon "doing it right the first time."

Common Threads of the Quality Thinkers

- Use ***facts*** rather than opinions
- Use ***data collection*** to gather information
- Use the ***scientific method*** to make decisions
- ***People*** are responsible for process improvement
- ***Systems*** and how people use them are responsible for quality
- Do it right the ***first time***

Frederick W. Taylor. Frederick W. Taylor is considered the "Father of Scientific Management." He promoted, through his actions and writings, the need to study, to theorize, to put theory into practice, and to revise theory and practice as necessary. Frederick Taylor was a practitioner and the first to see production management as a science. In 1919 he wrote in *Shop Management*, "Each man must learn how to give up his own particular way of doing things, adapt his methods to the many new standards, and grow accustomed to receiving and obeying directions covering details, large and small, which in the past have been left to his individual judgment." Taylor's practices led to workers doing what they are told and basically not being recognized for their ability. This leaves little personal judgment and truly invalidates the skills of the craftsperson and does not address group efforts. Today, the total quality management process states that people need to develop the procedures they use in their everyday work environment, and a team approach works best to accomplish this. Teams develop their own work procedures, rather than one individual or management. If press makeready is the issue, all the press operators work as a team to develop a makeready procedure that will work best for this group and the company.

Taylor was also an advocate of management by specialists. This meant that departments could have multiple managers depending upon their individual expertise. This created confusion and negated the proven process of having only one official leader. This concept has been modified to include staff personnel and team leaders, coaches, and facilitators. These positions carry the power of expertise and are not considered an "official" chain of command.

Taylor used an incentive plan that called for a daily accounting of each employee's output and the output's relationship to the earnings of the individual. In this way, he reasoned, the individual knew how he had performed and what it meant to his/her paycheck. He claimed that immediate feedback reduced dissatisfaction in what was required of the employee and the direct relationship to pay. Today, we are struggling with this concept. Taylor never discussed quality as a component of the process. He was focused on output and not the relationship between quantity, price, and quality. Today, we must make incentives a combination of these three components in a competitive printing marketplace. It is generally agreed that incentives work best when the individual can visualize what contribution was made to earn the incentive.

Taylor concluded that the general work force was basically incapable of planning because of its educational and technological illiteracy. Much of this was true at the time of his work but is no longer the case. It is not suggested that all people are capable of planning and are educationally and technologically literate. However, more people are capable of taking responsibility for their work and not only want, but demand, a say in how they do their work. It is therefore incumbent upon management to do a skills inventory of each employee's potential and build an organization that has its people trained in total quality management techniques. This skills inventory should include areas that show their capability for planning, education, and technological literacy and make a plan to assist each employee in developing those skills. (A sample skills inventory is shown on page 29.)

Walter A. Shewhart. Walter A. Shewhart is considered the "Father of Statistical Process Control." He was employed by Bell Systems (Bell Telephone Laboratories) and Western Electric (manufacturing arm of Bell) as a mathematician where, in 1924, he introduced the concept of statistical quality control. Bell Systems led the way in quality control by instituting an Inspection Engineering Department to deal with problems resulting from defects in their products.

He developed the Plan-Do-Check-Act cycle (PDCA), which consists of events that exist in any process improvement cycle. It is used as a methodology for process improvement

CASEY Press
Skills Inventory

Name ______________________________ Date ____________

Present Position ______________________________ Years Held ________

Past Position(s)/Number of Years ______________________________

Formal Education (completed) ______________________________

Skills

General (highest formal courses completed) ____________________

Math ______________________________

Science ______________________________

English/Writing/Speech ______________________________

Technical

Statistical Process Control ______________________________

Computer (hardware/software) ______________________________

Art/Design ______________________________

Prepress ______________________________

Press ______________________________

Bindery ______________________________

Managerial

Financial ______________________________

Human Resources ______________________________

Purchasing/Inventory ______________________________

Planning/Scheduling ______________________________

Supervisory ______________________________

Planning ______________________________

Total Quality Management ______________________________

Continuing Education (workshops, seminars, magazine subscriptions, memberships in printing organizations): ______________________________

____

Willingness and ability to accept new skills (managerial and technical): ________

____

for individuals as well as teams. The PDCA cycle is also a tool that should be used by everyone to accomplish their work.

Shewhart Cycle

- ***Plan***—What changes are desirable? What data is available? Are new observations needed? If yes, plan a procedure for gathering the information. Decide how to use the data.
- ***Do***—Carry out the plan, preferably on a small scale.
- ***Check***—Observe the effects of the change.
- ***Act***—Study the results. What did we learn? Implement, if desirable. Follow up to make sure the implementation is long lasting.

Shewhart is the developer of the control chart method of collecting data on a real-time basis. Kaoru Ishikawa, in his book *What Is Total Quality Control?*, states that "for the practical use of the concept of statistical distribution, the most convenient tools for this purpose are the three sigma control charts invented by Dr. W. A. Shewhart." Shewhart developed his theories and applications to chart the variations in a process and thereby control quality through analysis of the process. Shewhart claimed that when assignable causes are eliminated from the process everything remaining in the process is normal variation, inherent to the process and could not be removed without making basic changes in the process.

W. Edwards Deming. W. Edwards Deming worked summers, while doing graduate work, at the famous Hawthorne Plant of Western Electric Co. This is where he met his lifelong friend Dr. Walter Shewhart. After receiving a Ph.D. in physics from Yale University, he worked for the Department of Agriculture and the U.S. Bureau of Census. He was invited to Japan in 1947 to help with the development of statistical methods for the census of 1951. In 1950, he was invited back by the Japanese Union of Scientists and Engineers (JUSE). They were concerned with rebuilding Japan after World War II and producing quality products. In a series of lectures that he gave in Japan, he discussed the use of data collection, scientific methods, and statistical theory and principles to the management of quality. He explained that one could look at a sample of the population to infer and make decisions concerning the entire population. This statistical procedure was much quicker, less expensive, and just as accurate as counting the entire population. Deming claimed that if you believe in inspection of mass-produced products you cannot believe in statistics. People tend to make more wrong decisions because they do not understand or use statistical methods. He blamed managers for neglecting to give their employees better leadership toward the creation of better products. The Japanese honored Deming by naming their prestigious quality award in his honor and called it the Deming Prize for Quality.

Deming worked with NBC on a white paper entitled "If Japan Can, Why Can't We?" This documentary gave the American people an insight as to what statistical process control can mean to a country and an entire economy. Before the Japanese incorporated the use of statistical thinking into their manufacturing processes, they were known the world over as the makers of poor-quality products. Since instituting statistical thinking into their manufacturing processes, they have been known as a country that produces quality products.

Dr. Deming has worked with many companies in their quest for quality; in the United States, the most famous being Ford Motor Company and the General Motors Corporation. He emphasizes constancy of purpose for the improvement of products through the use and understanding of statistical analysis.

Much has been written about his "14 Points for Management." The fourteen points (shown on page 32), with minor clarification points in brackets, create a reference that will assist a company in meeting its quality process improvement mission.

Deming also believed in the use of the Plan-Do-Check-Act Cycle and promoted the Shewhart Cycle as a necessary tool for all quality practitioners. However, he calls it the Plan-Do-Study-Act Cycle.

Joseph M. Juran. Joseph M. Juran is an engineer who went to work for Western Electric at the famous Hawthorne Works in Chicago, where he was assigned to the Inspection Department. At that time there were 5,000 inspectors and a total of 40,000 employees.

He visited Japan in 1954 to assist them with their quality effort. Dr. Juran assisted the Japanese in changing their quality effort by helping them focus on the management

Deming's Fourteen Points for Management

1. Create constancy of purpose for improvement of products and services
2. Adopt the new philosophy [of refusing to allow defects]
3. Cease dependence on mass inspection [and increase reliance on statistical control]
4. End the practice of awarding business on the basis of price tag alone
5. Improve constantly and forever the system of production and service
6. Institute training [for all employees to meet customers' needs]
7. Adopt and institute leadership [supervisors need to be coaches and employees need to be given the proper tools to do the job right]
8. Drive out fear [encourage communication and productivity]
9. Break down barriers between staff areas [encourage different departments to work together]
10. Eliminate slogans, exhortations, and targets for the work force
11. Eliminate numerical quotas for management and people in the work force
12. Remove barriers that rob people of pride of workmanship [quality first]
13. Encourage education and self-improvement for everyone [provide ongoing retraining to keep pace with changing products, methods, and technology]
14. Take action to accomplish the transformation [clearly define top management's permanent, ongoing commitment to quality]

structure of business organizations. He did seminars for middle and top management. Before his visit, the Japanese were dealing with quality issues primarily as a technological process. Juran had come to the conclusion that there is a limit to the acceptance of statistical quality control when engineers are its prime mover. He focused his attention on the role and responsibility of top and middle management in the quest for quality.

Dr. Juran believes that quality issues need to be treated in the same way as financial issues. No company would think of being in business without organizing its financial structure. He believes that quality control is analogous to financial control, quality planning is analogous to financial planning, and quality improvement is analogous to cost reduction.

Juran suggested a project approach to quality improvement by identifying problems and scheduling them for solutions by ranking them in order of importance. He focused on determining which problem(s) would assist the most in improving the process. He is an advocate of Pareto analysis where problems are analyzed by "solution of the vital few" in which one can visualize through graphical representation that 80% of the printing problems are caused by 20% of the contributing factors. When narrowing down any problem, it is generally no more than two or three items that are causing 80% of the problem. Juran says that this is where the effort should be given and not to the "trivial many" contributing factors. In this way, the printing company can focus its energies on reducing its inefficiencies and work on process improvement.

Dr. Juran also discussed the responsibility of an individual in determining assignable causes of defects. He claims that each individual is responsible for approximately 15% of the variations in printing; the other 85% are conditions that are not under the control of the employee. A press operator is given the press, the paper, the ink, and generally the conditions under which to work. Maintenance time and money is a condition that is not under the control of the operator. The only controls, in this case, are how the operation will come together to make a printed product. These assignable causes are under the control of the professional, craftsperson, technician, or laborer. Operator carelessness, not following specifications, ignoring parts of a standard operating procedure, or being tired, either physically or mentally, are considered assignable causes and are approximately 15% of most print quality problems.

Management is responsible for 85% of manufacturing malfunctions. These are under the control and guidance of printing managers. These causes are the procedures, materials, and equipment chosen to do the printing job. Lack of preventive maintenance programs for equipment, lack of effective training programs, and the lack of adequate specifications and verification programs for incoming supplies are part of the malfunction of the process. It is a manager's responsibility that these causes be eliminated.

Juran has developed the "Journey from Symptom to Cause," which is a quality improvement action plan. This matrix is developed having a specific activity that can be a quality "steering arm" or a "diagnostic arm" activity. Typical quality steering arm activities and typical quality diagnostic arm activities are shown on page 34.

Typical Quality Steering Arm Activities

- Assigning priorities to project(s)
- Theorizing on the cause of symptom(s)
- Narrowing the list of theories
- Approving the design
- Providing authority
- Proposing remedies
- Taking action to institute the remedy
- Controlling the new level

Typical Quality Diagnostic Arm Activities

- Pareto analysis of symptoms
- Testing theories
- Collecting and analyzing data
- Designing experiment(s)
- Conducting experiments
- Establishing the proof of cause
- Testing the remedy

Juran has also offered a methodology for printing companies to use with their suppliers. He calls the methodology "self-certification" for product quality. This includes requesting that your suppliers have a failure prevention analysis and quality plan for their products. After the above plan has been agreed upon and a number of shipments shown to have quality compliance, the customer has to only audit occasionally to make sure quality remains stable. The customer conducts supplier visits to be sure the systems for quality are in place.

Dr. Juran suggests that one should study what customers do as well as what they say. Conformance to customer specifications is a moving target and will continue to be so. Customers naturally expect more from the product over a period of time. Products will continue to improve because customers not only demand but expect this to happen.

In the development of process improvement, Juran sees that self-directed work teams will replace the Taylor management process where there is a separation between those who plan and those who execute the plan. Self-directed work teams will plan and execute the work. This will reduce supervisory layers while making the team its own internal supplier and customer. This level of responsibility will combine the two forces of planning and execution of the plan.

Kaoru Ishikawa. Kaoru Ishikawa is the "number one" leader of Japan's quality resurgence. In 1949 he became involved with the Japanese Union of Scientists and Engineers (JUSE) in the promotion of quality. Dr. Ishikawa served as a full professor at the University of Tokyo.

Professor Ishikawa summarized his thinking in his book *What Is Total Quality Control: The Japanese Way*:

> As I look back on my life with QC, the following becomes my hope and prayer: That QC and QC circle activities be spread everywhere in the world, that quality all over the world be improved, that cost be lowered, that productivity be increased, that raw materials and energy be saved, that people all over the world be happy, and that the world prosper and be peaceful.

He combined the quantitative thinking of Juran, Deming, Feigenbaum, and Crosby, making it simple enough for factory workers to understand and use. He advocated understanding and using the seven tools of statistical process control (flow charts, Pareto analysis, cause-and-effect diagrams, check sheets, histograms, scatter diagrams, and control charts) to quantify manufacturing processes for understanding and improvement.

Dr. Ishikawa emphasized cause-and-effect diagrams (also known as fishbone analysis and Ishikawa diagrams), which take a close look at the effects of people, materials, equipment, environment, and methods on the process. His emphasis is on the practical applications of statistical tools. He moved away from sophisticated statistical methods because they sometimes made the understanding of quality problems more difficult for the average employee.

Professor Ishikawa differed from Feigenbaum, who advocated that total quality control be conducted by specialists and generally believes in rotating quality control managers as a stepping stone in management training rather than as a specialty unto itself. He believed that quality is a central part of management training and equal to accounting, production management, sales, and marketing training for managers.

Dr. Ishikawa believes that quality control begins with the interaction of people, and with this philosophy he emphasized the development of the quality control circle. Japanese businesses began using teamwork strategies in the early 1950s. This concept emphasizes that each work group, in an organization, will have an ongoing committee dedicated to finding and eliminating quality problems in their area of work.

Quality circles have not worked well in the United States, but teamwork practices are evolving from the basic quality circle problem-solving techniques that Ishikawa emphasized.

Armand V. Feigenbaum. Armand V. Feigenbaum was the first to use the term "total quality control." His approach aims at enlisting all parts of the organization in a quality effort including marketing, customer needs, willingness to put effort into quality, and manufacturing according to specifications.

Feigenbaum took a broader view of quality than his predecessors with his focus on the "9 Ms of quality."

Fundamental Factors Affecting Quality

- Markets
- Money
- Management
- People
- Motivation
- Materials
- Machines and mechanization
- Modern information methods
- Increasing product requirements

In addition to his Ph.D., Feigenbaum is a toolmaker, manufacturing executive, and founder of his own company. He also served in a top executive capacity with General Electric.

Dr. Feigenbaum defined the term "quality costs" as the cost of control and the cost of failure. He claims that companies need to track the cost of product failures (internal and external) both in doing and redoing the job. These costs tracking systems should include customer service, the measuring of quality, and quality development. The cost of control includes prevention and appraisal costs. Feigenbaum claims that these failure costs are a minimum of 10% and run as high as 40% of the annual sales dollar of a company. If the costs of quality are tracked and analyzed, a plan can be put into effect to eliminate these problems. This can mean as much as 40% to the bottom-line profits of a printing company.

Philip B. Crosby. Philip B. Crosby received a medical degree from Western Reserve University's Higher College of Pediatric Medicine. He found that medicine was not his calling and took a job with Crosley Corporation as a "reliability engineer."

Crosby soon became aware that he disagreed with what the quality professionals were doing, which was relying on inspecting completed goods and rejecting those that did not conform to specifications. He believed that management needed to realize that they had to produce products that did not fail. Quality had to be included as a major goal of the organization. The staff must receive training in how to achieve this goal.

Mr. Crosby was frustrated with statisticians who had no "perfect" in their thinking. The thinking of the time was that errors were inevitable. Quality programs were set up to predict how bad quality was going to get and adjust for this inevitability.

In an attempt to deal with product failure inevitability, Philip Crosby launched his "zero defects" program with the Pershing missile project at Martin Marietta Corporation. At that time, he was the company's director of quality. This program eliminated most of the Pershing failures, and "zero defects" became the official U.S. government policy for the defense procurement business.

The "zero defects" movement failed miserably, however. People misinterpreted the implementation of this program. Most companies that adopted the "zero defects" slogan failed to give it management commitment. Workers signed "zero defect" cards, and when the program failed it became the workers' fault. People were not taught how to do things right, but instead "zero defects" became a slogan and motivational approach without substance.

"Zero defects" is a powerful concept—one that is hard to conceive of as possible. One way to visualize the use of "zero defects" is to perceive of it as a process improvement strategy where one tries to achieve perfection in longer and longer intervals. If you can be perfect for a minute and keep extending that time frame until you are doing your job perfectly on a continual basis, then "zero defects" can become a reality.

"Zero defects" should be viewed as a developmental process much like weight lifting or any exercise where you build yourself up over time to the desired level. Doing your job in a quality manner is a "zero defects" strategy.

Crosby founded the Philip Crosby Associates Quality College where the four absolutes of quality are taught. These absolutes of quality are:

- Quality is defined as performance to requirements
- Preventing not inspecting quality out of the product
- Zero defects performance
- Measurement of quality is the price of nonconformance

Implementation of these four absolutes includes training targeted toward developing a new culture within an organization. To accomplish these absolutes, Crosby developed his fourteen points for the quality process.

Crosby's Fourteen Points for the Quality Process

- Management
- Quality improvement teams
- Quality measurement
- Cost of quality evaluation
- Awareness
- Corrective action
- Zero defects planning
- Quality education
- Zero defects day
- Goal setting
- Error cause removal
- Recognition
- Quality councils
- Do it all over again

The Crosby philosophy can be summarized with a quote from his famous book *Quality Is Free.* He states:

> Quality is free. It's not a gift, but it is free. What costs money are the nonquality things—all the actions that involve not doing jobs right the first time. Quality is not only free, it is an honest-to-everything profit maker... If you concentrate on making quality certain, you can probably increase your profit by an amount equal to 5 to 10 percent of your sales. That is a lot of money for free.

Crosby believes that quality is relatively easy to achieve once people care about it, measure it, and believe that defects are not inevitable.

Jenichi Taguchi. Jenichi Taguchi is a statistician who worked as an engineer for Nippon Telephone and Telegraph Company and is a teacher of quality control for the Japan Standards Association. The basic tenets of his philosophy and teachings center around the relationship between the quality of a manufactured product and the total loss created by the product to society when quality is not considered or implemented. This is known as the "Taguchi Quality Loss Function." The necessity of continuous quality improvement and cost reduction for an organization's health in a competitive economy and the need for never-ending reduction in variation in products and services are the basics of the Taguchi Quality Loss Function. Taguchi proposes that we utilize the design of experiments methodology. This methodology considers the best method of developing specifications and manufacturing processes. It looks at and suggests researching through experimentation to determine the best method of production.

He extended quality improvement to include product and process design as an off-line quality control system. Taguchi offers a method that provides a system which designs specifications into a product and/or process and produces a product that can continuously surpass these specifications.

Taguchi Off-Line Control System

- Quality is measured by the total loss to society
- Continuous quality improvement and cost reduction are imperative
- Quality improvement is a never-ending reduction in variation
- It is mathematically possible to determine society's loss due to quality malfunctions
- Product and process design significantly affect quality and price
- Reduce variation by determining desired performance characteristics
- Product and process variation can be reduced by utilizing statistically designed experiments

Without optimization of quality, there is customer dissatisfaction and higher service cost. Taguchi says we should design "robust products." A robust product is one that takes into consideration how the customer will abuse the product. In printing, this might include heavier covers on magazines, varnish to reduce scuff on annual reports, and shipping containers that are waterproof. In a society that is aware of the environment, a robust product may be considered one that considers reducing the paper, finding reuses of the product, and/or recycling the product.

Agreement between Current Theoreticians and Practitioners of the Quality Movement

In the United States, there are those who propose that we follow a specific quality guru (theorist). We would like to suggest that you study the theories of as many of the quality thinkers that you can. As described above, each of the great theoreticians and practitioners has made significant offerings that will assist us in reaching our quality mission. The following are some of the commonalties that are prevalent in present quality discussions and actions:

- Every organization needs a commitment to quality from top to bottom. We agree that top management must "buy into" goal setting for quality and support the quality goal by allocating time, people, and financial support.
- Companies must identify their most critical quality problems and management must lead in solving them. Management must lead through coaching and facilitating to obtain the best thinking of all members of the organization.
- In every process, graphic arts companies must try to define those characteristics that indicate quality so they will know what to change and how to measure success. This will also include service.
- Companies must achieve quality by understanding and improving systems and by preventing problems rather than by reducing defects by inspection and correction. We must work toward reducing fire fighting. Employees must be empowered to understand and assist in developing and maintaining specifications.
- Companies must develop a statistical understanding of processes and use statistics to solve problems. Statistics is defined as the collection, organization, and interpretation of data that describes the entire graphic arts process or product. Data collection must be random in order to assist us in determining whether the process is in or out of control. Statistics allows for accurate diagnoses of problems.
- When special causes are eliminated, problems in the process are generally inherent variables that need to be

controlled. When we understand what we need to control, we must continually work on improving the process.

- Total quality management is a requirement of a healthy printing company. This methodology requires that all aspects of the organization adapt the principles of statistical process control. Quality is not to be defined by product output alone, but by those elements that make up the entire organization including marketing, sales, manufacturing, and distribution.
- Quality can only happen through people. Employees must assist in developing those elements that are needed to define the printing organization's quality mission. Employees must be empowered to work with an organization, not for an organization. Management must facilitate these actions.
- Quality must be understood as a profit center with long-term, strategic implications.
- In order to prosper, printers must provide ever-improving value to their customers. Exceeding—not meeting—requirements should be the goal for developing and maintaining customers.

Summary

In this chapter, we have discussed how the printing industry, with the diverse types of products it manufactures, fits into the different manufacturing categories. Printed products vary from one-of-a-kind, to almost assembly-line manufacture, and these differences must be considered when addressing the various aspects that comprise a quality system. We have also discussed the dominant thinkers in the quality movement who spent their professional lives in theory and in action in order to make manufacturing more productive and responsive to the needs of the customer.

3 The Economics of Quality Improvement

Conventional Wisdom: What's Wrong with It?

It has often been said that there are three things that the customer desires from the printer: price, service, and quality. Conventional wisdom states that the customer, even if he/she wants all three, can only have two of these. The reasoning here is that better service and higher quality cost more. If customers want a low price, then they will have to give a little on service or quality. If they want high quality, then they will have to give a little on service or price. The major assumption is that price is the driving factor in determining quality. As is often the case with conventional wisdom, however, the logic is appealing but the assumptions are flawed. This chapter proposes that quality—not price—is the driving factor and that the ability to deliver a product at the lowest price is directly tied to the quality of the processes involved in printing the job.

The following diagram and the one on page 44 present the conceptual framework for this chapter.

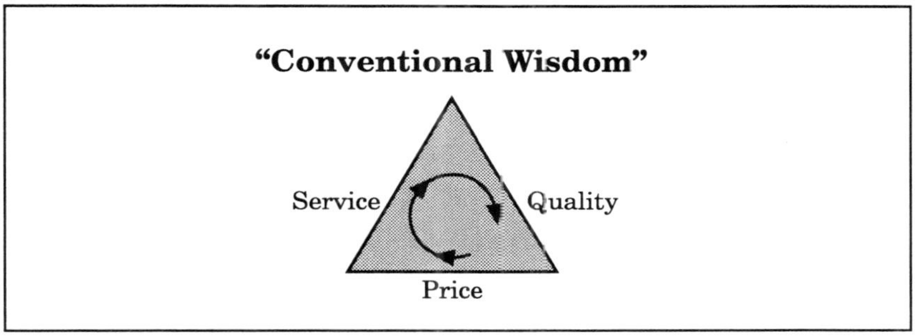

In the "Conventional Wisdom" diagram, quality and service are derived as a function of price. The customer who wants higher quality must pay a higher price. In other words, "price buys quality" or "you get what you pay for."

"Contemporary View"

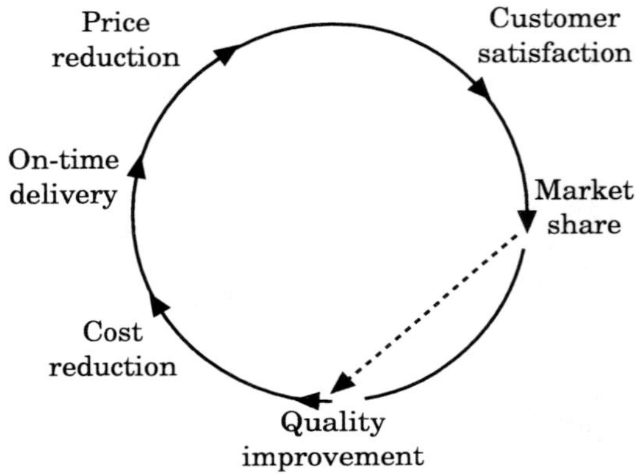

In the "Contemporary View" diagram, the quality of goods and services produced will determine the cost of the printed job, the ability to deliver on-time, the price charged for the job, the level of customer satisfaction, and ultimately the share of the market that an individual printer has. Actually, it's really just a question of how we define quality. If we define quality as the ability to deliver defect-free goods and services in a timely manner, then the "contemporary" view of the world is accurate. As the quality of a process increases, there is greater utilization of resources.

High-Quality Processes and Product Features

High-quality processes are synonymous with low waste and spoilage, few job reruns, little need for mass inspection, and higher customer retention rates. Quality improvement from this standpoint is a profit-making venture rather than another cost burden that the printer must bare. However, if we define quality in terms of features, the "conventional" view is more accurate. What is a feature? A feature is what we would generally use to refer to a different grade of product. Items such as stochastic screening, foil stamping, UV coating, special colors, Hi-Fi Color, and waterless printing are product features, and the purchaser will generally be willing to pay for them if they are desired. However, they are not synonymous with higher quality, which is really a question of consistency and minimal variations within a set grade of features.

What Are Quality Costs?

"Quality costs" are the sum total of the costs involved with making a product correctly. In printing there are two choices:

either the job is printed correctly the first time or it must be reprinted until it is correct. Quality costs are one of the best means for quantifying the overall level of quality, since they take into account the entire impact of both problems and improvements. Quality practitioners have broken down quality costs into four general categories:

1. ***Prevention costs:*** the costs involved with printing it right the first time. This includes items such as training and giving people the time and resources to do the job correctly.
2. ***Inspection and appraisal costs:*** the costs involved with verifying that the product meets specifications. This could include SPC measurements of in-process work, sampling of raw materials like paper and ink, or final inspection of product before sending it to the customers.
3. ***Internal failure costs:*** the internally incurred costs of not having printed the job right the first time. This includes any type of rework or waste incurred that wouldn't have been if the process functioned perfectly.
4. ***External failure costs:*** the customer-related costs of failing to print the job correctly. This includes costs such as customer turnover, excessive sales expense, and loss of market share.

Unfortunately, traditional methods of accounting and control have failed to look at these categories as separate elements on the income statement. Instead, quality costs have been lumped in with such general items as labor, materials, overhead, and selling expense. Recent findings in companies like Motorola and TRW have shown that, when separately accounted for, quality costs can be as high as 20–30% of sales revenues for manufacturing organizations. In service organizations, the figure can reach as high as 50% of sales.

Why Should You Measure Quality Costs?

There are several reasons why you will want to start measuring your quality costs. These include:

1. Determining the *return on investment* (ROI) for quality improvement—when you spend time and effort on improving a process, what is the return realized?
2. Justifying individual quality improvement projects—to what lengths does it make financial sense to go to improve a process?

3. Benchmarking the overall impact of the quality improvement process—what is the net effect of your efforts over time?
4. Getting top management attention—dollars are the language of management. Cost of quality allows you to put quality at the level of something meaningful to managers.
5. Giving direction to your improvement efforts—quality costs, by focusing in on waste, rework, and customer satisfaction, can get the whole company moving in the same direction.

In the words of Dr. Joseph Juran, quality costs represent "...gold in the mine." This is money that the printer is already spending. As companies invest resources in their quality improvement processes, the managers of those companies will want to see the return on that investment. If the company decides to spend $100,000 a year to train employees in SPC methods, then looking for the return in the form of reduced need for inspection, reduced rework, and reduced customer turnover will allow management to assess the profitability of that investment. It is very seldom that a company does something because it's "the right thing to do." More often than not, companies will improve quality either because the customer demands it or because they understand that there are significant financial rewards involved. Tracking quality costs allows managers to make the right decisions in regards to allocating resources to the quality improvement process.

On a project-by-project basis, it is important to look at quality improvement activities in a cost/benefit framework. The final decision to go ahead with a quality improvement project should be made on the basis of that project's financial merit. Estimating the quality costs involved will allow quality teams to evaluate the viability of their proposals. If a quality proposal involves investing $30,000 in training activities, but the expected return is only $10,000, the project should be scrapped. Just as we look at the decision to purchase a new press, step-and-repeat machine, or stitcher-collator from the standpoint of expected ROI, so too should we look at quality improvement projects from a financial standpoint. Sometimes, however, benefits are intangible or immeasurable, especially when it comes to customer satisfaction. In such cases, a project might be worth pursuing even though it can't be "proven" immediately in financial terms.

As a printing company gets involved in quality improvement, it will begin to see a multitude of benefits. Productivity will increase. Scheduling will become more efficient. Customers will be more satisfied. While all of these are desirable, it will be difficult to assess the overall process by looking at these individual elements. Quality costs provide a means of evaluating the overall impact of the process. In essence, quality costs are the common denominator by which all of the benefits can be measured. Tracking quality costs over time will give the company the ability to see that it is improving over time. This is especially true since most benefits will be incremental in nature. If the total of incremental improvements is tracked, the company will be able to more objectively set quality improvement objectives. Rather than saying that it desires to reduce spoilage by 20% or that it wants to increase productivity by 15%, an objective like "reducing failure costs by 10%" would take both of these into account.

Quality Cost Categories

In this section, we will look at a more detailed breakdown of these categories of quality cost (prevention, inspection and appraisal, internal failure, and external failure). Each of these costs will be evaluated in terms of its avoidability in reducing the total of all quality costs.

Prevention costs. Prevention costs represent in large part the investment that the printing company will make in quality improvement. Traditionally, prevention has had a very low priority in the United States. These costs represent the up-front time and effort required to do the job correctly the first time. Typical prevention costs include training, preventive maintenance, vendor certification, planning, and quality team meetings. Prevention costs are wholly unavoidable if the printing company is to reduce its overall cost of quality. In other words, prevention costs are the price a printing company has to pay for real quality improvement.

Inspection and appraisal costs. Inspection and appraisal costs represent all of the various ways in which we look at the product to ensure its conformance to requirements. This process starts and ends at the shipping and receiving dock and takes place at various checkpoints throughout the process. Instead of preventing problems from occurring in the first place, many printers will look at the product and inspect the substandard pieces out. Inspection and appraisal costs are partly avoidable and partly unavoidable. As quality levels increase, the need to inspect finished products will be reduced. However, true quality improvement involves allowing employees such as platemakers and press operators to appraise their products in order to control their processes. These costs are unavoidable, but they will also be diminished as processes come into a more stable, controlled state. An excellent example of an appraisal cost in the printing industry is color proofing: it is generally considered a necessary process, although by standardizing your reproduction process from computer monitors to imagesetters and consistently fingerprinting your equipment as well as getting color separations tailor-made for your presses, it is possible to minimize the use of color proofs for monitoring purposes.

Internal failure costs. Internal failure costs represent what happens when we haven't done the job right the first time. Some of the printed matter will be thrown away, some will be reprinted, valuable prepress and press time will be used for reprinting the job, and additional inspectors will be required to make sure that the defective product doesn't reach the customer. Internal failure costs are wholly avoidable when the proper preventive measures have taken place and in-process inspections have ensured product conformity. If you think about all of the waste and rework in your systems, you can get some idea of the potential savings that can be realized by focusing on internal failure costs.

External failure costs. External failure costs represent what happens when the customer gets defective products. These costs include liability costs, claims and discounts, and high customer turnover. It has been found that an unhappy customer will typically tell five to seven of his/her friends about the problems associated with the printer involved. Think about how you react in your personal life when you go to a restaurant and receive lukewarm soup. If you are anything like a print buyer, you will tell your friends about the experience and never go back. What's worse, customers don't usually complain, they just leave. If quality improvement is to be effective, printers must ask customers not only what they liked about the job, but what they didn't like as well. External failure costs are also wholly avoidable. When the printed product is delivered on-time and defect-free, the customer will react favorably and be retained. Your sales effort can then focus on truly new customers, not just replacing those that have left because they were dissatisfied.

Relationship between the Cost Elements and Strategies for Cost Reduction

The following diagram shows the relationship between the various quality cost elements and the overall cost of quality. The model was originally proposed by Dr. Joseph Juran and shows how the cost elements interact with one another.

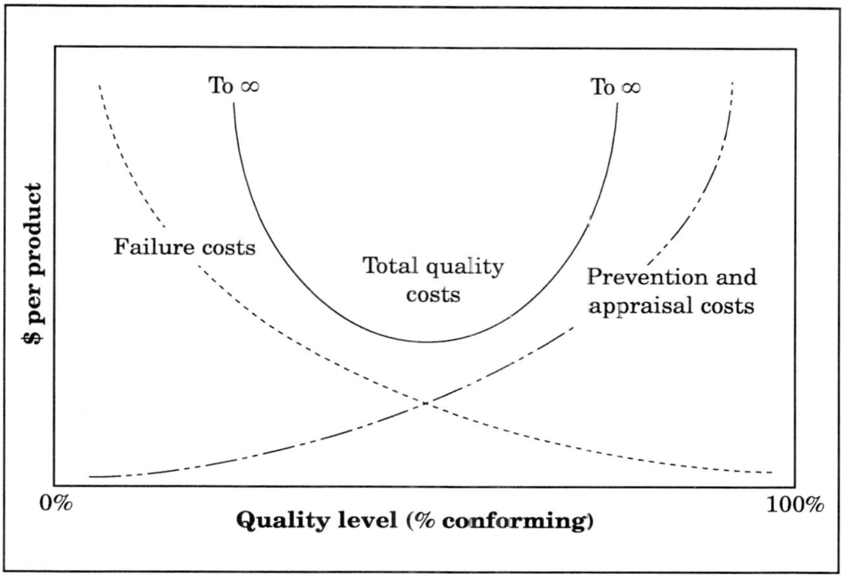

This diagram shows that at low levels of prevention and appraisal the result is excessively high levels of failure costs. The overall cost of quality is thus very high. However, as prevention and appraisal costs increase (the company invests in quality improvement and quality control), failure costs begin to decrease rapidly. As a result, the total cost of quality is reduced. This process continues to a point beyond which any increase in prevention will be more costly than the resulting return in reduced failure. One should be wary in determining when this break-even level has been reached. Since many failure costs (such as customer dissatisfaction) are intangible, it is difficult to accurately determine the true magnitude of external failure costs. Therefore, relying solely on tangible quality costs as a reason for declaring the "war" to be won is probably a mistake.

The levels of prevention, inspection and appraisal, internal failure, and external failure represent strategic choices made by the printing company's top management. A company can choose to try and "inspect quality in" by having inspectors and quality control people, or a company can choose to focus its efforts on improving quality and preventing problems. The following hypothetical model demonstrates each of these choices and the cost effects associated with a particular strategy.

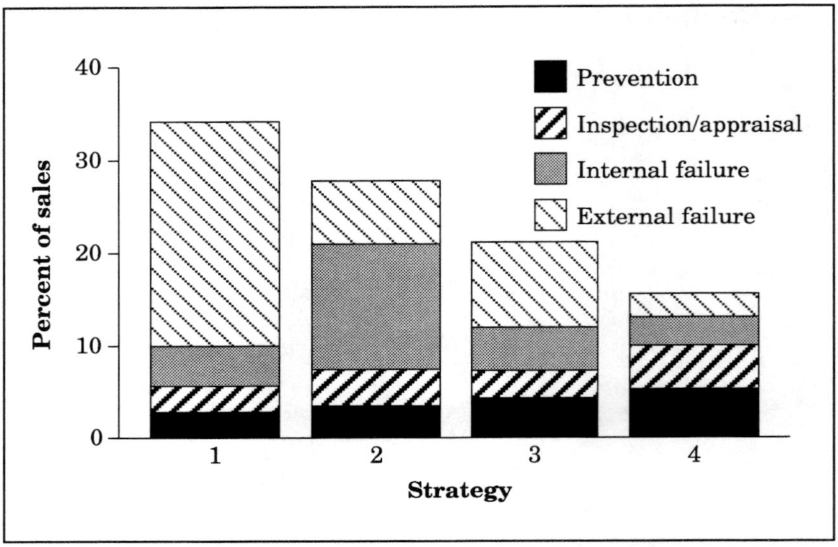

These four strategies could be labeled as follows:

- Strategy 1: "let's wing it and hope for the best strategy"
- Strategy 2: "sort and suffer strategy"
- Strategy 3: "prevention strategy"
- Strategy 4: "balanced prevention and inspection/appraisal strategy"

In the "let's wing it and hope for the best" strategy, the printer has decided that he/she is too busy for preventive maintenance, SPC, training, measurement, specifications, and vendor certification. As a result there is some internal failure cost involved with waste and spoilage and low productivity. However, internal failure is fairly low for the simple fact that mistakes and defects aren't discovered because nobody's checking. External failure costs are extremely high, as customers never return, and if they do it is to say that they want a discount or are filing a liability suit. To get new customers, the printer must employ a large sales force. This is an extreme case.

The "sort and suffer" strategy is more typical of traditional printers. The bulk of the quality investment is made in "QC people" to look at the product (proof inspection, plate checkers). This company doesn't have a quality improvement process, but rather a quality control department to fight fires and inspect products. Since the quality of the printing process isn't improved but the product is being looked at, this strategy yields very significant internal failure costs. Many press sheets are being thrown away. This company is in the practice of printing a certain percentage of "overs" and accepts high spoilage and waste rates. On the upside, external failure costs are reduced by this strategy, since less substandard product gets through to the customer. The overall result is a lower overall cost of quality than in the first case, but still much too high.

In the "prevention strategy," the printer has decided to invest 5% of sales in preventive activities. While this is a step in the right direction, external failure costs are still quite high. Note that internal failure costs are less than in the "sort and suffer" mode since the printer has opted not to inspect the quality of the product. In this case, the print buyer is the inspector of the product.

In the final approach, the printer is dedicating fully 10% of sales to preventive activities and in-process inspections. As a result, failure costs (internal and external) are reduced to virtually nothing. This could also be classified as the mature phase of quality improvement. This strategy yields the lowest level of total quality cost. Assuming that the printer started in the second strategy, the return on investment realized by moving to the fourth strategy would be measured as the increase in prevention and inspection/appraisal divided by the reduction in total quality cost. In this case, the printer increased prevention/inspection/appraisal from 7% to 10% and saw a reduction in total cost from 27% to 14%. Thus:

$$(10-7)/(27-14) \times 100 \approx 23\%$$

This is a healthy return by anybody's criteria. Furthermore, the reduction of total quality cost means that 13% of sales is going back to the bottom line. If the company was grossing $1,000,000 and netting $100,000 before changing strategies, its return would have been 10%. After changing strategies, it would be netting the original $100,000 plus an additional 13% of $1,000,000. Thus, the company would now be netting $230,000, or 23%! This has the same effect as if the company had increased its gross sales to $2,300,000, without actually having to increase sales a dime! Quality improvement is a real opportunity to make money.

The important thing to remember when focusing on overall quality cost reduction is that prevention generally yields the greatest return on investment.

Implementing a Quality Cost Measurement System

In order to start measuring the costs of quality, it is recommended that the printer use the following steps:

1. Get top management support.
2. Find a quality costs "guru."
3. Select and assemble the quality costs team.
4. Select a pilot area for implementation.
5. Establish the pilot program.
6. Widen the program to encompass the whole organization.
7. Track costs over time and use the information to steer the quality improvement process.
8. Teach teams to use cost/benefit analysis.

Top management support. As with all undertakings, it is essential that the boss is committed to the process. Senior management must understand the benefits of tracking quality costs as a mechanism for improving quality and strengthening the bottom line. Top management must allocate the time and resources necessary to put the program in place and be committed to following through on the information provided by the ongoing analysis. If at this point, senior management doesn't see the quality costs program as being as important as other budget items, it would be a fruitless exercise to continue. On the other hand, if you can afford to put in place a very limited program without much support, this can be a good way of generating better support for an overall program.

Quality costs guru. Someone in the organization must be found who is willing to champion the process. This person will serve as the catalyst for the members of the quality costs team. The guru should have the following characteristics:

1. Strong commitment to the program.
2. Respect in the organization, from top to bottom.
3. Good leadership and team skills.
4. An understanding and concern for financial matters.
5. Perspective for all quality costs.

Good candidates for a quality costs champion would include the chief financial officer, the controller or estimator, or anybody else who has the above qualifications. Be careful not to select someone who will be over his/her head or will let the process get bogged down in minutiae. The person must also be willing to "invent" the system, not being constrained by the present financial accounting practices.

The quality costs team. Establishing the quality costs program will require input from a variety of sources. These include finance, marketing and sales, purchasing, quality control, and production. The team's job is to establish and implement the system for collecting, processing, and analyzing the cost data that will be produced. The team will need to have an understanding of materials, labor, and machine costs; specifications; work flow; customer needs; vendor capabilities; and internal customer/supplier relations.

Select a pilot area for implementation. The quality costs team should select a pilot area for implementing the program. The purpose of a pilot implementation is to learn how to put a system together and work out the kinks before going "live" with the project. The pilot area should have the following characteristics:

1. High impact on the bottom line and the customer—what happens if a mistake is made in this area?
2. Low level of resistance—don't try to implement your quality costs program in a department that is headed up by the most obstinate supervisor in the plant.
3. Easily quantifiable—the department should have a fairly straightforward process, with tangible inputs and outputs.
4. Manageable scope.

The easiest way to select a pilot area is to first break down the printing process into functional areas: art and design, shipping and receiving, prepress operations, pressroom, and finishing operations. The team should ask itself how each of these areas rates in terms of the above characteristics. As part of this process, it would be helpful to talk with sales, marketing, and customer service to assess the importance to the final customer. The financial people will be helpful in looking at the potential impact on the bottom line. Interviewing personnel from the various functions will give the team a feel for the resistance, quantifiability, and scope of the operation. A final ranking of the areas should give a clear answer as to which area best serves the purpose of a pilot program.

Establish the pilot program. Once an area has been designated as the pilot, the team should select a few key personnel from that area to join the team in implementing the pilot program. The team must then begin to develop a checklist of the quality costs within that area. Brainstorming, flow charting, and cause-and-effect analysis are useful tools in this process. A list of all quality costs must then be broken down into the categories of prevention, inspection and appraisal, internal failure, and external failure. Next, the team will come up with a methodology for collecting data on each of these cost areas.

For prevention costs, the team should look at how much time and resources go into making sure things happen correctly the first time. What is the dollar value of training employees? What is the dollar value of giving teams the time to investigate projects? Are resources being used to improve the quality of incoming supplies and materials? How much is being spent on preventive maintenance? Are time and materials being used to calibrate processes such as plate exposures?

For appraisal costs, the team should look at how much money is going into verifying product quality. This includes the cost of process control and measurement instrumentation and the time to take the measurements. Also, any inspectors in the department would be included in this list.

Internal failure costs pertain to the costs incurred by that department when it has to do something over. For example, if plates have to be remade because of dirt and dust in the vacuum frame, the team would count the time to remake the plate times two, since the platemaker would normally be doing productive work instead of remaking the plate, which is nonproductive. Also, the costs of direct materials used in the plate remake need to be included. In this process, the easiest thing to do is to multiply the time-related cost elements (budget hour rates, labor, administrative overhead) times two and then add to that the direct materials and the indirect materials/supplies charges. However, only do so if you are convinced that the operation is supplanting a productive activity. Also, if the remakes are taking place in overtime, or regular jobs are being done in overtime because remakes are being done in normal time, be sure to take the additional labor cost into account when multiplying times two. This point deviates from that made in other cost of quality accounting texts, which make no such "times two" suggestion. This is strictly the opinion of the authors of this text, who wish to take production opportunity costs into account when measuring internal failure.

Measuring external failure costs will depend on where the defect is detected. In this case, the 1-10-100 rule is applicable. A plate remake in the platemaking department might cost $10. If the defect is caught in the pressroom, the cost might be $100 since the pressroom incurs lost productivity as a result. If caught in the bindery, the cost might be $1,000 since the job has been printed and must be reprinted—in addition to lost productivity in the bindery. If the customer finds the defect, the cost might be $10,000 since the job will need to be totally redone and the customer might ask for a discount in addition to telling everyone that he/she knows that your printing is lousy. The point here is not the exact numbers, but the principle that quality defects are like avalanches that increase in scope exponentially as they get further and further from their source.

When the team begins looking at all of these costs, it is best to stick only to the hard numbers, such as press downtime and paper waste, rather than trying to quantify such things as cost of customer turnover. While the intangible costs are certainly important, any measurement of these will be suspect. Those individuals in the organization who wish to torpedo or cast doubt on the validity of the numbers will have ammunition to do so if the numbers aren't hard. In any case, the team probably will have its hands full collecting all of the hard data in the pilot area.

Widening the quality costs program. Once the pilot program is functioning and has the bugs ironed out, the next step is to begin expanding the quality costs program to the other functional areas of the organization. The team should proceed on a function-by-function basis, using the same criteria that it used for selecting the pilot area. The implementation in each area should follow the same basic steps as in the pilot area, with the exception that it isn't necessary to have the program fully in place in one area before proceeding to the next. If prepress, press, or bindery costs are being investigated, a member from that department should be included on the team.

Track costs over time. The collected data should be formatted so that the team can track the behavior of the individual cost elements over time. This information can be used to assess the nature of the quality improvement process and to steer management in the direction of meaningful projects. Probably the best means for tracking quality costs over time is to use a run chart at first and later a control chart such as an individuals chart or a moving averages chart (see Chapter 7) to look at cost elements as a percentage of such bases as net sales or value added. A typical run chart for quality costs might appear as shown below.

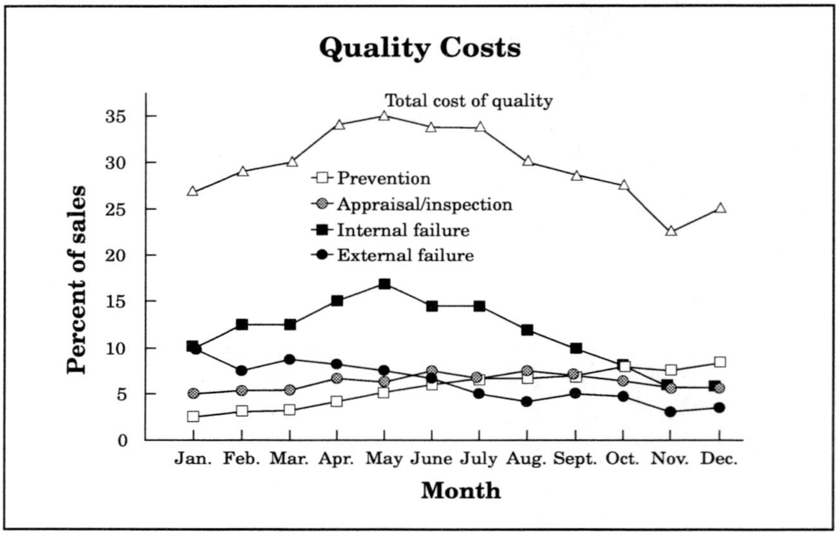

This chart illustrates what typically will occur when getting into the quality improvement process. Notice the trends in the cost elements over time. Prevention costs increase from day one, as do inspection and appraisal costs, which then fall slowly over time as the need declines. Internal failures at first skyrocket due to increased inspection and a lag in the impact of prevention activities. External failures fall from the beginning and continue throughout. The overall impact is a total cost of quality that at first increases and then rapidly decreases thereafter. Thus, before getting into continuous improvement, printers must be aware that an initial decline in productivity will result. This is roughly analogous to the learning curve adjustment process experienced when a new piece of equipment, such as a new step-and-repeat machine, is installed. The implication is that senior management must be ready to ride out the storm before the benefits begin making themselves known. If you are not prepared to stick it out, it is well advised not to get involved in the first place, since the results of abandoning the process can be very demoralizing for all involved.

Cost/benefit analysis. Quality cost data can be useful in showing potential projects for teams to address. These teams must be educated on the nature of quality costs and use this information in formulating their proposals for quality improvement projects. Teams should look at the additional costs that will be incurred by the project such as startup (learning curve) costs of implementation, additional or incremental prevention and inspection/appraisal costs involved with the project, capital outlays, and training costs. The benefits are evaluated in terms of reductions in internal and external failure costs and overall improved productivity. It is important to project these costs and benefits for a reasonable period of time and then discount them back to the present using conventional valuation methods. Many printing companies look at a payback period method in making capital investments. This is a suitable alternative to use in the cost/benefit analysis.

Pareto analysis is a useful tool for teams in collecting data for project justification. The team should prioritize the various projects by the dollar value involved in their implementation. For example a company could use this methodology to solve its platemaking problems by first looking at the various causes of platemaking defects and then evaluating these by the dollars typically associated with each type. Finally, the company can implement a solution (such as an air filtration system for dirt and dust), which may pay for itself in less than six months.

Quality Cost Worksheet

When you get started in your quality costing process, a worksheet is useful. Your quality cost team and champion can use the worksheet on page 58 as a tool for improving their effectiveness.

Summary

The discussion of the economics of quality improvement in a printing company has demonstrated that quality improvement is an excellent opportunity for printers to make more money. The categories of quality cost allow printers to break down and track quality costs so that they may stay on track

Quality Cost Questionnaire

1. What will be the purposes(s) of the quality cost program?

2. Who will champion the program? Why?

3. Who will be on the program development team? Why?

4. Where will the pilot program be implemented? Why?

5. What will be the cost elements?

   Prevention

   Appraisal

   Internal failure

   External failure

6. How will costs be collected and reported?

7. Who will be responsible for these tasks?

and see the benefits of their efforts. Many printers won't improve quality because "it's the right thing to do," but they will do it to increase profits. The first step in this journey is to recognize that quality doesn't cost, it pays! It is recommended that the printer apply a systematic team-based approach to putting in place a quality costing system, as this is a large and time-consuming project. The wise quality change agent will use quality costs as a means for guiding the improvement process in a fashion that is based on data and focuses the whole company on customer satisfaction.

4 Data Collection and Analysis

Management by Facts and Data

The use of facts developed from quantified information can help people to avoid making decisions from hunches, hearsay, or emotion. This should not preclude listening to people who have experience and a "gut feeling" about the printing process. These experiences and gut feelings must be quantified and verified before action is taken. Experienced employees need to be coached in the use of statistical process control tools and the scientific method so that their vast experience can assist in determining those areas in which facts need to be collected. It is very important to verify gut feelings.

Manage with Facts and Data

- Collect data.
- Analyze with data.
- Speak with data.
- Act with data.

Comments such as "the press can not hold register" or "the folder will not fold accurately" need to be verified through quantified data collection. How accurately the press holds register or how far off the fold is on the signature must be quantified. Are register problems consistent or intermittent? Are you selling jobs that you can not print on your present equipment? Maybe you need to look for work that has looser specifications? Maybe your setup procedures are incorrect or your maintenance program needs improvement. It also may be that a minor tune-up from the equipment manufacturer will solve quality problems. These questions can not be answered by hunches, hearsay, emotions, and gut feelings.

Although experience is valuable, it might not be applicable or sufficient to a given situation. Past experience might color what the person "sees" as fact. Experience should be used to identify potential problem areas and solutions. Past solutions might not lead to process improvement. Sometimes, previous solutions were bandages that covered over the real problem, making it harder to resolve in a permanent manner.

Non-TQM Reactions to Problem Solving

- Hunches
- Hearsay
- Emotions
- Gut feelings

The use of facts and data give people within a printing environment a powerful means to successfully solve problems and improve quality. Facts and data are not just the purview of managers, but are the heart of problem solving and process improvement for all members of the organization. Managers must learn and lead the way in the use of facts and data.

Key Points in Data Collection

Revealing and/or verifying the existence of a problem are basic reasons for collecting data. Without facts it is difficult, if not impossible, to determine whether what you are doing, be it service or manufacture, is consistent and in statistical control. The use of facts can tell you whether you have a problem in the diverse areas of the printing business. Does your company have problems with time management of the sales force or the accuracy of estimates and on-time deliveries? Without data collection, these questions can not be accurately

Reasons for Collecting Data

- To ***reveal*** or ***verify*** the existence of a problem
- To ***analyze*** a problem
- To ***prevent*** a problem
- To verify to a ***customer***

answered. Data will give you the information to determine if a problem exists. It is possible to believe that you have a problem in a specific area of your organization only to find out that this is really not your problem. Converting the numbers collected into pictorial explanations of Pareto diagrams, scatter diagrams, histograms, and control charts will assist in determining the importance of the data and the relationship between the variables.

Analyze a problem without data, and the charting and graphing of that information makes it virtually impossible to analyze what is happening within a given operation. Using the tools of SPC from the data that you collect will enable you to analyze the relative importance of the information. Quantifying the possible reasons for customer loss or retention will assist you in developing strategies for retaining customers.

Using the data you collect and using the tools of SPC will assist your organization to be proactive in preventing problems. Data and its application will give you the advantage of seeing where you are going and allow you to head off problems while improving your company. If the data collected assists you in detecting a high percentage of jobs being shipped late, you may wish to take action to remedy this problem before you lose customers. Finding solutions to problems are opportunities for preventing future problems.

Verifying to a customer that you can produce a printing product to agreed-upon specifications through the use of data collection and its analysis assist in verifying the value-added aspects of a printed product to the customer. Charting density, register, and print-to-fold characteristics will not only impress your customer, but add value to your claims of being a quality printer.

Collecting the right data is an essential part of the process of a quality printing organization. It is essential to formulate a clear statement of what the potential problems are and to define what is to be measured. It is necessary to determine which aspects of the printing process from the sale of the job to its delivery are the most important aspects of quality and then determine how these can be quantified through data collection.

The development of data collection forms that keep the purpose of the process in mind and focus upon collecting only relevant data is essential to a well conceived and run printing company. The form must be kept simple with time-efficient "form friendliness" as a central focus. The entire organization must be considered when developing data collection forms. How a particular form fits into the entire total quality management process needs to be addressed. When forms are used as an end unto themselves, they can cause a breakdown in data collection and its many quality uses. Before adding a new form to the quality process, the printing organization must continually question its relevancy.

Question the Data Collection Form

- What is the purpose of the form?
- How much time does it take to fill out the form?
- How will the form be used?
- Who will fill out the form?
- Does the form replace a present form?
- Can the form replace a present form?
- How will the form improve our printing company?
- How does the form affect the quality delivered to the customer?

Necessary Precautions in the Measurement Process

It is important to work through people. People are sensitive about their errors. They must know that data collection and measurement of the process are tools that will assist them to do a better quality job. People must be allocating the appropriate time to taking the measurements. Data collection and measurement can not be used to judge the operators. Emphasis must be placed on evaluation of the process. Control needs to be in the operators' hands, and they need to know that they are responsible for quality. When people make serious mistakes, the most likely reasons are the tools they use, the education and training that they bring to the job, and how they are managed. Management must be cognizant that people need to be provided with the proper tools, trained, and coached in how to put it all together to produce consistent quality printing.

Measure only those areas that are most important to the end-product quality or service; it is not necessary to measure everything. Do not burn people out with measuring items that do not conform to the basic considerations of what constitutes the critical customer considerations of quality. Overmeasuring can become a detriment to efficient printing operations. Data collection needs to be part of the quality process and not a detriment to the operation. Be careful not to overburden operators with insignificant measurement and data collection. Measurements need to be precise and timely. On-line control needs to be done so that operations can be adjusted as the printing proceeds. This procedure will reduce and eventually eliminate the rejection of out-of-specification work. The primary objective is to do it right the first time.

Be sure the measuring device is precise enough to allow for the accuracy required to ensure conformance to specification. The wrong measuring tool or the lack of calibration can invalidate the results of process control measurements. If the print-to-diecut registration is $\pm\frac{1}{64}$th of an inch, be sure that the measurement device is capable of this precision. Proper measurement techniques and analysis are a critical component of statistical process control and continuous process improvement.

Management Commitment to Data Collection

1. Managers must learn to collect, analyze, and display data using the appropriate tools of statistical process control.
2. Once managers understand the need for and the uses of data collection and analysis, they must educate everyone in the use and display of data.
3. Managers must insist that decisions be made through the use of appropriate data.
4. Managers must coach employees in the use of the scientific method for problem solving. Employees must go to the actual source of the problem, see the actual problem, talk to those directly involved, and use data for verification of the problem.

Data Collection Principles

- Go to the actual source of problem.
- See the actual problem.
- Talk to those directly involved.
- Use data to verify the problem.

5. Displaying the data is the heart of control and improvement. Data must be made available to everyone so that it can be analyzed and used in problem resolution.

6. Management must ensure that there are clear understandings of quality attributes at each step of the process.
7. Management must demand that data be supplied for all quality concerns. Data should be considered suspect if it can not be verified.
8. Management must ensure that those people involved in specific operations actually collect and plot their own data in a prescribed, timely manner.
9. Management must insist on the use of data for determining performance of all aspects of the printing process from sales to customer satisfaction, from customer service to order entry, and from manufacturing to delivery.

Measurement of Critical Print Variables

What variables to measure. It is essential to identify those variables that cause quality problems. The best way to identify these variables is through customer and supplier input. Understanding the needs of your customer is the primary focus for identifying critical variables. Most customers do not know or need to know specific ink, paper, or other printing specifications. They will generally make their choice of paper and ink and assume that each and every copy will be the same. This is not the case. The making of paper, ink, and the printing process itself are dynamic and will change from copy to copy. Knowing that this is true is the first step in developing specifications that take into consideration the normal variations of each process and the needs and expectations of the customer. Customers are generally not interested in or knowledgeable of the process variation. They do know what they will and will not accept as a quality product. The customer does know how far the ink color or the whiteness of the paper can vary before it affects his/her perception of quality. It is important for the printers to understand how accurately their suppliers can hold those specifications that are important to the customer. The printer must communicate customer needs and expectations to the supplier of paper, ink, and other incoming raw materials. It is incumbent upon the printer to require that suppliers have quantified specifications for the variables that help determine end-product quality. Suppliers must have a procedure in place for monitoring these variables. The printer needs to work as an intermediary between suppliers and customers so that all parties fully understand what are the specifications and the acceptable variation.

A common complaint in the printing industry is color density variation. You may wish to work with customers in determining density ranges for their printing jobs. You do not have to discuss the theories of densitometry with the customer, but you need to determine what the customer considers acceptable variation from the given specification of color density. It is then the printer's responsibility to set up a procedure to maintain this specification. This involves working with the ink maker to determine acceptable pigmentation of the ink and the control of the variation of this pigment level. Explaining to the customer the need to know his/her tolerance range and how you will maintain it also develops partnerships in quality performance.

What to measure. The first step in measuring variables is to take a broad view of those areas that affect the quality of printing. After determining that the process is in statistical control, specifications need to be developed for those areas that tend to vary and need to be monitored.

Some Areas to Measure

- Incoming raw materials
- Prepress
- Press
- Finishing
- Shipping

The Process of Determining Measurement Procedures

The use of the form on page 68 will assist you in setting up an effective variable measurement system.

What do we want to measure? Measure those items that are critical to the outcome of the printing job. These can be partially determined by the materials and equipment used. The customers' needs will also determine what to monitor. Work with suppliers and customers to identify variables that will make the end product a success.

Why do we want to measure it? Measure those variables that will determine the success of the printed job. It is very important to understand why you want to measure a variable. Not all variables have the same impact on the end-product

Specification Development Form

1. What do we want to measure? ______________________________

2. Why do we want to measure it? (main concerns) ______________________________

3. What device(s) are available for measurement? ______________________________

4. What is the specification? ______________________________

5. What is the tolerance? ______________________________

6. Who will do the measurement? ______________________________

7. Does the person(s) understand the importance of the measurement? How was this determined? ______________________________

8. Is the person(s) trained on how to take the measurement? ______________________________

9. How often is the measurement required to be taken? ______________________________

10. How is the measurement recorded? ______________________________

11. What happens to the information after it is recorded? ______________________________

quality. If the size of the film is a little larger or smaller, this variable may not be of any consequence to the printed job. However, the density of the film may have a great effect upon the end product. Therefore, it is important not to spend a great deal of time and energy measuring the film size, but time must be spent measuring density of the film. Concentrate on determining which variables are the main concern to the end-product quality and concentrate on quantifying and controlling them. If supplier ink tends to vary in hue, then it is important to not only have hue specifications but a system to determine if the strength is being maintained.

What device(s) are available for measurement? After determining the need for a specification and how to quantify variation within the specification, it is necessary to determine what instrument to use to measure the variation. Developing a supplier conformation report is one method of working with suppliers. The quality of the supply is monitored by the supplier, and the statistical information is then given to the printer for verification. Sometimes it is a matter of using a simple go/no-go gauge. Determining the proper distance of staple holes from drill holes, when a particular job requires this specification, can be done with a simple gauge, e.g., a small gauge to determine the distance between staples and drill holes. At other times it may be training a person to view color under specified lighting conditions. This can be done with a high-low color sample and a 5000 Kelvin light booth. Sometimes it may require more expensive electronic devices. Specific color density may require the use of a hand-held densitometer or even a scanning densitometer.

What is the specification? Determining the specification generally begins with the customer. The printer must know what the customer will accept in the final product that is produced. This is generally dealt with when the salesperson takes the order. It is important that all specifications be understood before starting the printing process. However, the customer does not have to know all the specifications that will go into the printed job. The opacity of the paper or the contrast level of the printing plate are not specifications that most customers are involved with. These specifications are derived from what the final product needs to look like. Incoming materials have specifications that are derived from industry standards and supplier capability plus the needs of the printer as determined by the customer. Specifications also come from the printer's capability. Not all printing presses are capable of holding the same register or dot gain. When the printer understands the equipment capability, it is possible to develop a specification that can be explained to the customer and agreed upon as a quality goal for each printing job. There are generally accepted specifications that are industry standards. Specifications for Web Offset Publications (SWOP) have specifications for type of paper, ink hue, grayness, density ranges, and dot gain that most publication printers agree is the standard specification for their printing work. This methodology has been accepted by some commercial printers as a guide to specification development.

What is the tolerance? When you are dealing with variables, specifications need to have tolerances expressed in plus or minus values. These tolerances need to be understood by all parties involved in the printing process. The supplier as well as the printer and customer must realize the effects of variation upon the completed printed product. Some accepted tolerances include how many under or overs a printer can deliver in final production and still be in compliance with customer's expectations. This tolerance has generally been agreed to be ±10%. As long as the printer stays within these tolerances, the end delivered product will be acceptable to the client. Of course, if a printer can deliver a printed product that is closer to the desired quantity, he/she may become the preferred supplier. This is known as exceeding customer's expectations. The objective is to reduce the tolerance through process improvement techniques. It is advantageous for suppliers and printers to understand the capabilities of what the supplier can produce and the needs of the printer in terms of what the customer expects as a quality product. When a printer receives out-of-tolerance supplies and attempts to "make do" by using these materials as part of the process, the end product will suffer. The end result is a cumulative effect dependent upon all the pieces that go into making it up. Too many printers accept out-of-tolerance supplies and then struggle in an attempt to make them work. This can be remedied by setting up specifications with tolerances and then requiring suppliers to adhere to them.

Who will do the measurement? Once a specification and its tolerances as well as how it will be measured are determined, it is essential to delegate the responsibility of the actual measurement to the proper person. This should be done by first looking at who is best qualified to do this measurement, whether this is the most cost-effective procedure, and will the measurement be done in a timely manner. For incoming supplies a printer may wish to have the supplier do the measurements and supply the data along with the materials. In many cases this is the most cost-effective procedure. Most printers do not have elaborate testing laboratories to check the many variables for such items as paper and ink. The agreed-upon quality data derived from suppliers needs to be reviewed by the printer. Therefore, it is very important to delegate this responsibility as part of someone's job description. Verification of supplies is critical to the quality of the printed job. Before starting a new lot of plates or film, most printers will do a calibration test to be sure that the exposure time has not changed. This is generally part of the platemaker's and camera operator's job descriptions. Generally, it is best to delegate the responsibility for these quality checkpoints to the person most involved in the process.

Does the person(s) understand the importance of the measurement? How was this determined? It is not only important to delegate the responsibility to an individual, but it is most important for that person to understand the importance of the assignment. If the individual feels it is just another assignment added to an already busy workload, the chances are that it may not get done. Even if it is done, it may be done erratically or haphazardly. In either case, the quality mission of verifying quality in a quantified manner will be defeated. It is not only important to explain the importance of quality verification but to give the task to someone whose job is partially dependent upon the product being in conformance with the specification. The person(s) who have the measurement responsibility should also understand the theory behind the measurement that makes their work important. This will assist in making sure that the measurement will be taken properly and continuously.

Is the person(s) trained on how to take the measurement? It is important to train the person responsible for the required measurement task. This can be done through supervisors doing the training or inviting suppliers to do the training. Do not assume that people know how to take the proper readings or when to take them. Most people tend to trust readings that are stamped on the supplies as accurate. If the form roller manufacturer claims that the roller is 25 A durometer, there is a tendency not to verify this measurement. If the blanket is stamped 0.065 in. (1.65 mm), there is a likelihood that we will accept this thickness as true. This has proven to be a problem, and therefore it is important to train your press personnel to take readings at a prescribed time, using a prescribed method, and record and report the data in a specific manner. When new people enter the work force, it is important to train them. Sometimes we forget to train new employees on how to verify important specifications, which will end up causing quality defects.

How often is the measurement required to be taken? It is not only important to have specifications and trained personnel, but it is also important to have a prescribed time sequence for taking each measurement. With expendable materials such as film, plates, paper, and ink, the verification may be by job or during the job. For more permanent supplies such as rollers and blankets, the time sequence may be by installation and after a given amount of time. Suppliers can assist in helping you set up a procedure for determining how often to take readings. Take readings as a means of controlling the process. If ink density varies every 10–15 minutes, a check should be taken every 8–12 minutes. If rollers get hard every six months, a printer may wish to do a check every five months. The time sequence used should be somewhat shorter than the cycle time of failure. Do not set the time sequence so close to the failure time that there will be a chance for failure before rectifying the out-of-specification condition.

How is the measurement recorded? Recording the measurement is very important. It not only alerts everyone to the fact that a check has been made, but recording is a good first step in process improvement. We can take information and graph it to see trends. We can use this information to explain to our suppliers what is taking place in the printing operation because of their process capability and control. We can also use this information to alert our customers as to what methods we use to ensure quality of their product.

What happens to the information after it is recorded? Many printing companies have measurement procedures

Specification Verification Form

Roller Type and Specification (±)	Readings				
	New	3 Months	6 Months	9 Months	12 Months
Form roller _____ ± _____					
Oscillator roller _____ ± _____					
Transfer roller _____ ± _____					
Date to be taken*					
Date taken					
Taken by					

*Fill in this row when the roller is installed.

that are very good, but either do not follow through on these procedures or do nothing with the report. This is time consuming and expensive. It also leaves employees with the feeling that they are doing work for nothing. The information that is recorded must be reviewed on an ongoing basis. Are we getting so much variation in the process (out of specification) as to consider changing a supplier? If the printer finds that every measurement is within specification, it may warrant taking fewer readings. This will save time and money. Sometimes, it is obvious that we no longer need to verify the specification. All that may be necessary is to develop a simple monitoring system.

The goal is to improve quality by developing quality specifications, verifying that they are being maintained, and working toward reducing variation. When we can quantify those elements that make up our quality mission and maintain those specifications, we can claim to be a quality printer and prove it.

Examples of Critical Quality Variables

The examples on page 74 show how a printing company can go about developing a specific program for its own quality needs. These examples include incoming materials, prepress, and press variables.

Measurement Procedure

Concern: ______________________

A. Instrument: ______________________

B. Main Concerns: ______________________

C. Quantification: ______________________

A sample of a paper variable

Concern: Paper Caliper

A. Instrument: Bench Micrometer

B. Main Concerns: Unevenness of Print; Feeder and Finishing Difficulties

C. Quantification: Mils

A sample of an ink variable

Concern: Ink Hue Error (deviation from primary color)

A. Instrument: Proofing Kit; Densitometer

B. Main Concerns: Color Image Reproduction

C. Quantification: Densitometric

A sample of a prepress variable

Concern: Contact and Vacuum Frame Control

A. Variables: Light Source; Drawdown; Materials

B. Methods: Step-off; Gray Scales; Resolution Targets

C. Evaluation: Visual and Densitometric

A sample of a press variable

Concern: Press Fingerprinting

A. Instrument: GATF Test Form/Reflection Densitometer

B. Main Concerns: Process Control and Capability

C. Evaluation: Densitometric

100% Inspection

100% inspection is not a statistical process. It is, however, used in printing where critical items can cause the entire process to fail. These items include estimates, sales orders, copy preparation, type specifications and proofreading, production orders, and customer specifications. 100% inspection is also used where critical products such as money, lottery tickets, or medical supplies are printed.

Sometimes, it is necessary to inspect up to five and six times to verify accuracy. Copy preparation, proofreading, and critical processes can fall into this category of inspection. Glass bottling is an example where this type of inspection system is employed because of the serious nature of loose glass in the bottle. One must be careful in setting up a system that truly inspects multiple times specified. Each inspection must be considered the most important one. It is easy to develop a mind-set that "if I do not catch an error someone else will." This defeats the purpose of multiple inspections. Each inspection has to be considered the final inspection and must be thorough.

When using the 100% inspection method, it is necessary to ensure that the inspection personnel do not get physically and mentally fatigued. The inspector must not be rushed in the inspection process. Rotation of inspectors and automated quality detection devices are two methods that should be considered to guarantee that the quality inspection level desired is actually maintained.

100% inspection does not always guarantee accuracy. The exercise on page 76 has been developed to assist you in realizing that 100% inspection does not guarantee that the end results are perfect. Keep in mind that there are time constraints when we inspect. There are environmental conditions (e.g., noise, people talking and asking questions, and other distractions) that make it even more difficult to concentrate on the inspection process. This example offers proof that careful scrutiny of some items can leave errors in the result.

100% sampling is a very exacting task that requires focused concentration, an amenable environment, time, proper training, and the correct tools.

Statistical Sampling

Economy and effectiveness are the chief advantages of acceptance sampling. As seen from above and the exercise on the next page, 100% inspection does not guarantee that there will be 100% conformance to specification and that all errors

Exercise: Count the "Fs"

Directions for the following paragraph: Count all the "Fs" and "fs" in the paragraph. Read the paragraph only once. Record your answer.

> The fatherly and famous handling of apprentices by Ben Franklin for effective and faithful work on the far reaching printing documents of our forefathers has been efficiently and forthwith promoted. Many felt that this was foolish folly for Franklin to follow. Careful financial efforts of a few of the fabulously wealthy fathers and mothers of the present printing community have seen financial fitness come to this fantastic endeavor. However, in view of the fairly rapid falling off of efficiency on the part of some of the more unfortunate of these family type printing facilities, it is feared that fairly soon the careful fundamental financial management of many of the family facilities may fall by the wayside. This would cause unfair rumors to be considered facts and cause fantastic harm to the faithful work of Ben Franklin.

Record the number of "Fs" and "fs" you counted.

Directions for the following paragraph: Count all the underlined "Fs" and " fs". Read only once. Record your answer.

> The fatherly and famous handling of apprentices by Ben Franklin for effective and faithful work on the far reaching printing documents of our forefathers has been efficiently and forthwith promoted. Many felt that this was foolish folly for Franklin to follow. Careful financial efforts of a few of the fabulously wealthy fathers and mothers of the present printing community have seen financial fitness come to this fantastic endeavor. However, in view of the fairly rapid falling off of efficiency on the part of some of the more unfortunate of these family type printing facilities, it is feared that fairly soon the careful fundamental financial management of many of the family facilities may fall by the wayside. This would cause unfair rumors to be considered facts and cause fantastic harm to the faithful work of Ben Franklin.

The correct answer is 66 "Fs" and "fs." Were you correct? Chances are that you were not.

will be found. Therefore, statistical sampling offers printers an opportunity to verify the accuracy of work in an economical and quantifiable manner.

For employees, statistical sampling is less boring and less time-consuming than inspection as only relatively small samples are inspected. Reviewing the quality of each manufactured item can become a laborious task and lead to mental and physical fatigue. This can also lead to inaccuracies in the quality inspection process.

Where the item to be inspected is expensive and has to be destroyed to validate its quality level, the statistical sampling process is necessitated. If you are quantifying the specification of glue adhesion on the spine of a perfect-bound book, it is not practical to destroy every book to validate this specification. You will want to set up a statistical procedure that verifies that the glue is adequate to provide reliable results, while destroying as few books as possible.

Failure to statistically measure what you plan to measure. It is not only important to have statistical process control procedures and specifications developed, but it is imperative to follow these procedures. Having a sampling procedure in place is a major step in quality printing, but following these procedures makes the system work. If a press operator does not pull the appropriate random press sheets, the collecting of specified print characteristics is rendered statistically invalid. Failure to follow procedures for sampling and measuring incoming materials can lead to poor-quality print outcomes. Sampling and measurement need to be done in a timely, prescribed manner to be effective.

Measuring devices. Errors due to inaccurate measuring devices are a common occurrence. Using quality measuring devices that have not been calibrated properly or often enough will yield erroneous figures. A printing company that does not follow an agreed-upon procedure for calibrating its densitometers will inevitably have readings that do not correspond to reality.

Calibration devices. Calibration devices need to be checked regularly for accuracy. Using a durometer gauge which has been calibrated to a rubber plaque that is old may give you a durometer calibration reading that is not accurate. Rubber gets hard with time, and so the rubber plaque may have a hardness that is not what the manufacturer has given as the original hardness. It is important to have calibration devices checked for accuracy on a regular basis. Devices that have not been checked for accuracy can also yield numbers that are erroneous and misleading. Follow the manufacturer's prescribed procedure for checking the calibration device. If the manufacturer has not given you a procedure for checking the accuracy of the calibration device, you need to contact the manufacturer to get this important information. Keep a record of the calibration information so that it can be updated in a timely manner.

Operator error due to carelessness. Carelessness can come from a lack of understanding of the need to perform to specifications. It is important for everyone to understand why the measurement of quality is important. If possible, allow those people involved in measuring quality to discuss and assist in specification development. Encourage customers to discuss the importance of their specification with manufacturing personnel.

Operator error due to fatigue. Employees who are fatigued will tend to allow quality control to slip. People who do not get enough rest, exercise, or proper nutrition become fatigued more readily than those who lead a balanced life style. Excessive work hours will tend to reduce an employee's ability to make sound quality judgments. It is in an organization's best quality interest to offer programs that encourage people to live a balanced life style. Printing companies that have work schedules of 10- and 12-hour shifts need to be aware that employees will get more tired than those companies that follow an 8-hour work schedule. Developing quality systems that are followed will assist in mitigating the fatigue factor.

Operator error due to inadequate instruction. Having the correct measuring device for quality is not the only prerequisite for a quality outcome. People who use the device must be trained in how to use it, how often to use it, how to calibrate it, and how to record the numerical data that is received from the quality control device.

Transcription errors. Once the quality device is used, the information must be recorded. It is easy for people to transpose numbers, leave numbers out, put information in the wrong place, and generally be less than 100% accurate in the transcription process. Whenever possible, the quality control device should have a "closed loop" operation. Having a densitometer directly linked to a computer relieves the operator from the possible chance for error. Using a wand to input data rather than having an individual input data by hand reduces the chance for human error. Wherever possible, use a system that uses few numbers and letters.

Errors involving the use of computing equipment. Computers are machines and as such are as suspect to accuracy as any other device. The adage "garbage-in, garbage-out" must be remembered. Wrong keys can be struck by an operator causing the results to be tainted with inaccuracies. Software might contain glitches, or the user may not know how to properly use a certain software program. Computing equipment has made data collection and the use of statistical process control much easier; however, use of the proper programs and equipment is mandatory in order to gather the appropriate quality information.

Attribute measurements. An attribute is defined as something either being present or absent. Missing staples in a book, hickeys on a litho-printed sheet, and typographical errors are printing attributes because they are characteristics that can be measured as being present or absent. An attribute can be classified as *yes,* a certain characteristic is present, or *no,* a certain characteristic is missing. Go/no-go gauges are considered attribute measuring devices.

Variability measurements. In nature, all outputs of a process vary. Sometimes the variation is so insignificant to the customer that we do not consider it relevant and therefore claim that each specimen is perfect. Quality that is measured in plus or minus (±) is a variable. Variables are measured on a continuum. Some common items that are measured as variables are register, dot gain, density, and paper thickness.

Sampling frequency. The frequency of sampling depends upon the printing process's tendency to vary over time. Printing presses and folders tend to cycle in their variation more frequently than rapid-access processors and light-integrated vacuum frames and therefore need a sampling plan that accommodates these variations. A sampling plan has to be incorporated into all operations and adhered to in order to assist in maintaining and verifying those characteristics that are designated to represent quality of the printed product. Run charts and control charts can assist in determining the sampling frequency that must be used to ensure quality.

Number of samples. To develop statistically reliable information, a minimum of 30 samples are needed to adequately develop an SPC charting system. If possible, 100 samples should be gathered and used in the charting process. This may be impractical for short-run applications in the printing industry. However, one may consider grouping samples according to the similarity of many short-run jobs as a means of obtaining statistically valid sizes of 30 or more samples. If you are running the same type of signature on a folder, it is possible to group various jobs together so that they all can be considered one job. This will allow you to gather more samples throughout the combined run. Remember, the more samples the better!

Subgroup sampling. In most processes there is a tendency to vary from sample to sample. Therefore, the use of subgroup sampling is common. A typical procedure would be to take 3–5 samples in a row and average them to come up with one averaged observation for that particular sample. Subgroup sampling is also required when using a control chart that incorporates averages and ranges.

Observation by time. Each process should be scrutinized for its own SPC needs. Some areas to consider include the machine's ability to hold settings accurately, the variation of the materials to be used, the importance of the operation in terms of variation, and whether the equipment runs continually or sporadically. A program then can be set up to determine the sampling frequency.

Observation by percentage of operation. It is important to use sampling plans that are practical. If the press or bindery operator is required to take samples at a rate of one sample every 1 or 2 minutes, there is a great likelihood of these procedures not being followed. Operators are generally very busy and will not have the time to thoroughly inspect a printed or bound signature every 1 or 2 minutes. However, it is very important to use acceptable sampling procedures. Do not use arbitrary sampling schemes. Have operators examine only the most critical customer specifications on a minute-to-minute basis and examine less critical specifications in a looser but statistical manner. Operators must save all samples for quality assurance purposes. Do not put them back into production until they have been thoroughly reviewed. These saved samples can be used for further random sampling and/or assisting in developing subgroups.

Common Examples of Quality Control Procedures

- Light-integrated vacuum frame should be calibrated weekly.
- Litho plates should be monitored using a GATF Plate Control Target or an UGRA Plate Control Wedge on each plate.
- Film should be calibrated for each film batch change.
- Monitor the printing process every 10–15 minutes. Use shorter durations of time for shorter runs and higher speeds of web litho or when the process is most likely to change rapidly, i.e., beginning of a run, shift changes, material changes (roll or skid change).
- Monitor the bindery folders/stitchers/collators/trimmers every 10–15 minutes. Use shorter durations of time for shorter runs.
- Monitor incoming supplies upon the arrival of each shipment.

The following sample size procedures have been used successfully by some printers.

Practical Sample Size Procedures

- After the OK to run is given, sampling is at a rate of two percent (2%) up to a run of 5,000 (two inspection samples for every 100).
- One percent (1%) sampling for runs between 5,000 and 100,000 (one inspection sheet for every 100).
- One-half percent (0.5%) sampling for over 100,000 (example: for a run of 200,000, inspect one product for every 200).
- If similar short-run jobs are run back to back, they can be lumped together into one longer run, and use the higher run length for sampling procedures.

Statistical Sampling Plans for Acceptance Sampling

Acceptance sampling by attributes. The following acceptance sampling plan can be used for print suppliers (internal and external) and customers to determine whether to accept or reject a specific order. This sampling system is for go/no-go situations; e.g., the number of samples out of specification for register, color, staples missing, or glue not holding. This methodology can also be used for sampling incoming materials (e.g., paper and ink), operations, or services (e.g., number of times a telephone rings before it is picked up or how many times specifications are missing on an order). This sampling process can also be used between departments to assure proper quality and for final inspection before shipment to the customer.

The most practical system to use for acceptance sampling is MIL-STD-105E (Military Standards), issued on May 10, 1989 by the United States Department of Defense. The use of this methodology has also been adopted by the American National Standards Institute and the American Society for Quality Control. An example of acceptance sampling using this methodology is given on pages 84–85.

Objectives and uses of attribute acceptance sampling using MIL-STD-105E. The objective of using statistically sound principles in attribute acceptance sampling is to assist in protecting the customer and supplier from shipping or receiving poor-quality products. Acceptable Quality Level (AQL) sampling inspection plans, as determined by the customer and agreed upon by the supplier, are a means of assisting in quantifying this process. It reduces the risk of arbitrary rejection, as the procedure is statistically derived and agreed upon by both the supplier and customer.

Some applications for the printing industry include the development of vendor rating systems, vendor certification, attribute acceptance sampling within the printing process, final inspection before shipping, and final print customer product acceptance.

MIL-STD-105E is a method of attribute acceptance sampling that has been developed by the United States Department of Defense and is widely accepted by industry as an appropriate procedure for attribute sampling.

One must be aware that there is some risk involved in this methodology since this method of inspection is based upon sampling and is therefore used with a limited sample size. However, the risk is statistically known and calculated on what is called an operating characteristic curve. One can use this risk as the basis for selecting a particular plan.

Military Standard Sampling Procedures and Tables for Inspection by Attributes (MIL-STD-105E) and Military Standard Sampling Procedures and Tables for Inspection by Variables (MIL-STD-114) can be obtained from Naval Publications and Forms Center, 5801 Tabor Avenue, Philadelphia,

PA 19120. A simple slide rule combining the attribute tables entitled AQL *Inspection Manual and Rule* is available from INFO, P.O. Box 606, Ayer, MA 01432.

Basic MIL-STD-105E terminology. To use the MIL-STD-105E methodology for attribute acceptance sampling, one must first understand basic terms in acceptance sampling procedures.

Acceptable quality level (AQL) is a designated maximum level of percent defective (defects per hundred units) for which lots will be accepted or rejected. The AQL level is the chance the printer and customer will accept for the sampling plan chosen being wrong. There are 26 AQL levels; however, the most common are those that are within the range of 0.065–15, or 13 levels. In the printing industry, the most common AQLs that are chosen are between 1.0 and 2.5. These AQLs give the printer and customer a 97.5 to 99% confidence level of making a correct decision given the sampling procedure. The AQL is a reflection of the chances that the vendor, printer, or customer wishes to take that the sampling plan chosen will yield the required results. If 1.0 is chosen for the AQL and we are inspecting staples in a booklet, it is assumed that 1 book per 100 may have a missing staple before the product is rejected.

Acceptance number is the maximum number of defects or defective units in the sample that still allow for acceptance of the lot represented by the sample.

Rejection number is the minimum number of defects or defective units in the sample that will cause rejection of the lot represented by the sample.

A ***defect*** is any nonconformance to specified requirements. If the spine of a booklet is specified to have three staples and only has two, this would be considered a defect.

Inspection is the process of measuring, examining, testing, or otherwise comparing the unit of product with the requirements.

Lot or ***batch*** is the entire shipment or quantity from which a sample size is to be drawn.

A ***sample*** is a unit that is drawn from a lot at random.

The ***sample size*** is the number of units taken from a batch.

Sampling size code letter is a device used along with the AQL for locating a sampling plan on a table of sampling plans.

Sampling plan indicates the number of units of product from each lot that are to be inspected and the criteria for determining the acceptability of the lot. The plan includes

Example of Acceptance Attribute Sampling Methods Using MIL-STD-105E

The following is an example of how MIL-STD-105E can be used in a printing business. It is strongly suggested that a printing company, its suppliers, and customers purchase the entire MIL-STD-105E manual. This will enable the reader to understand the details and develop a broader perspective of the potential of this methodology.

Facts:

1. 2,700 posters are to printed as a four-color job on a four-color litho sheetfed press
2. The defect to be inspected is register and will be subjectively viewed with a 12-power magnifying glass. The specification is ±½ row of dots. The screen used for this job is 150 lines per inch.

Question 1: What is the sample size when MIL-STD-105E, for attribute acceptance sampling, is used for this job?

Solution:

1. Use the *Sample Size Code Letters Chart* on page 85.
2. Scan down the left-hand column until you come to 1,201 to 3,200 (2,700 posters is between these two numbers).
3. Moving from left to right, find the Sample Size Code Letter "K" under General Inspection Level "II" (normal inspection).
4. Go to the *Acceptable Quality Level Chart* (on page 85) and scan down the left-hand column (Sample Size Code Letter) until you find the letter "K."
5. The first column to the right is the Sample Size and reads 125.

Answer: 125 samples must to be taken randomly from the 2,700 printed posters.

Question 2: Based upon the rules of MIL-STD-105E, how many samples can be out of specification before the job is rejected. The customer has requested an AQL of 2.5.

Solution:

1. Scan from left to right across the top of the *Acceptable Quality Level Chart* until you find AQL of 2.5.
2. Go to the column entitled Sample Size Code Letter until you find K (as determined from question 1). To the right of this column is the Sample Size column, which for Sample Size Code Letter K is 125.
3. Go across from K and 125 until it crosses the 2.5 AQL column. You will read the Ac (accept) and Rc (reject) numbers of 7 and 8.

Answer: Seven (7) samples can be out of specification. The job will be rejected if 8 or more samples out of 125 are found to be out of specification.

Sample Size Code Letters Chart

Lot or batch size	General Inspection Levels I	II	III
2–8	A	A	B
9–15	A	B	C
16–25	B	C	D
26–50	C	D	E
51–90	C	E	F
91–150	D	F	G
151–280	E	G	H
281–500	F	H	J
501–1,200	G	J	K
1,201–3,200	H	K	L
3,201–10,000	J	L	M
10,001–35,000	K	M	N
35,001–150,000	L	N	P
150,0001–500,000	M	P	Q
500,0001 and over	N	Q	R

Acceptable Quality Level Chart

Sample size code letter	Sample size	Acceptable quality levels (normal inspection) 0.010	0.015	0.025	0.040	0.065	0.10	0.15	0.25	0.40	0.65	1.0	1.5	2.5	4.0	6.5	10	15
		Ac Rc	Ac Rc	Ac Rc	Ac Rc	Ac Rc	Ac Rc	Ac Rc	Ac Rc	Ac Rc	Ac Rc	Ac Rc	Ac Rc	Ac Rc	Ac Rc	Ac Rc	Ac Rc	Ac Rc
A	2	↓	↓	↓	↓	↓	↓	↓	↓	↓	↓	↓	↓	↓	↓	0 1	↓	↓
B	3	↓	↓	↓	↓	↓	↓	↓	↓	↓	↓	↓	↓	↓	0 1	↑	↓	1 2
C	5	↓	↓	↓	↓	↓	↓	↓	↓	↓	↓	↓	↓	0 1	↑	↓	1 2	2 3
D	8	↓	↓	↓	↓	↓	↓	↓	↓	↓	↓	↓	0 1	↑	↓	1 2	2 3	3 4
E	13	↓	↓	↓	↓	↓	↓	↓	↓	↓	↓	0 1	↑	↓	1 2	2 3	3 4	5 6
F	20	↓	↓	↓	↓	↓	↓	↓	↓	↓	0 1	↑	↓	1 2	2 3	3 4	5 6	7 8
G	32	↓	↓	↓	↓	↓	↓	↓	↓	0 1	↑	↓	1 2	2 3	3 4	5 6	7 8	10 11
H	50	↓	↓	↓	↓	↓	↓	↓	0 1	↑	↓	1 2	2 3	3 4	5 6	7 8	10 11	14 15
J	80	↓	↓	↓	↓	↓	↓	0 1	↑	↓	1 2	2 3	3 4	5 6	7 8	10 11	14 15	21 22
K	125	↓	↓	↓	↓	↓	0 1	↑	↓	1 2	2 3	3 4	5 6	7 8	10 11	14 15	21 22	↑
L	200	↓	↓	↓	↓	0 1	↑	↓	1 2	2 3	3 4	5 6	7 8	10 11	14 15	21 22	↑	↑
M	315	↓	↓	↓	0 1	↑	↓	1 2	2 3	3 4	5 6	7 8	10 11	14 15	21 22	↑	↑	↑
N	500	↓	↓	0 1	↑	↓	1 2	2 3	3 4	5 6	7 8	10 11	14 15	21 22	↑	↑	↑	↑
P	800	↓	0 1	↑	↓	1 2	2 3	3 4	5 6	7 8	10 11	14 15	21 22	↑	↑	↑	↑	↑
Q	1,250	0 1	↑	↓	1 2	2 3	3 4	5 6	7 8	10 11	14 15	21 22	↑	↑	↑	↑	↑	↑
R	2,000	↑	↑	1 2	2 3	3 4	5 6	7 8	10 11	14 15	21 22	↑	↑	↑	↑	↑	↑	↑

how many samples are to be taken, AQL, and defect accept or reject number.

Normal (II), tightened (III), loosened inspection (I) are included in MIL-STD-105E specifications. Use normal levels (II) of inspection and refer to the MIL-STD-105E manual for changing this procedure.

Destructive sampling procedures are included in MIL-STD-105E. These procedures are useful when it is necessary to take relatively small samples because one must destroy the sample to determine the attribute; i.e., tear apart an expensive perfect-bound book in order to determine whether the binding will hold up.

Cost of Sampling

It should be noted that the same job with reduced inspection levels needs a random sample of 50 (Sample Size Code Letter H at Level I). This is preferred supplier status and requires that at least 10 runs have been accepted before going to this plan. A reduction in sampling means that less cost is incurred in the sampling process. The cost of quality, as it pertains to inspection, is a benefit as less time has to be spent inspecting (125 for normal inspection and only 50 samples for reduced inspection). If one can assume that 1 minute is spent on each sample for inspection purposes, it will be necessary to spend 125 minutes for normal inspection and 50 minutes for reduced inspection to determine conformance to the registration specification. If the person doing the inspection is paid $10 per hour or 16.6¢ per minute, the cost for normal inspection is $20.75, and $8.30 for reduced inspection. This is a cost savings of $12.45. This is the savings for acceptance sampling for a relatively small job. One can readily determine that preferred customer status is very profitable. A TQM printing company can be more profitable by doing the job right the first time, thus minimizing the need for sampling inspection.

Summary

In this chapter we have discussed the need for management to commit to the use of facts and data in the operation of the company. The appropriate information needs to be collected in a scientific manner so that the organization can monitor quality and analyze process improvement strategies. Data collection systems need to be developed and used to manage the quality of the printing process. By asking and answering the appropriate questions, the organization will be able to develop specifications, train employees to gather and analyze data, and use this information to meet customer demands.

5 The Problem-Solving Tool Kit

An important part of total quality is the application of a structured process for problem solving. Whenever a team (or an individual) sets out to solve a problem, it can draw upon the tools and process described in this chapter to help ensure that it is going in the right direction and gets to the root of the problem. This process consists of the following:

- A framework for problem solving
- A tool kit for problem solving

In this chapter, the framework and tools are discussed as they relate to the overall context of continuous improvement. There are also several chapters (Chapters 6, 7, and 8) in which advanced techniques and applications of the problem-solving tools will be discussed.

Framework for Problem Solving

When attempting to solve problems, it is better to take a planned, structured approach than to just "jump right in." Several advantages can be given for a planned, structured approach, including:

- You will be more likely to get to the root cause of the problem (not just looking at the symptoms).
- You will be more likely to actually find a solution to the problem.
- Your solution will be more likely to "stick" permanently.
- Since you are more likely to actually solve your problem permanently, problem solving is more economical when utilizing a planned, structured approach.

The Plan-Do-Check-Act Cycle

One approach that has proven useful in solving problems for continuous improvement is the ***Plan-Do-Check-Act (PDCA) cycle.*** PDCA is a process for identifying, analyzing, and solving problems in a way that is modeled on the scientific method, but which works well in the business and printing environment. The four phases of the cycle are*:

1. ***P—Plan:*** the problem is defined, data is collected about the problem, and several potential solutions (hypotheses) are developed that might help solve the problem.
2. ***D—Do:*** the hypotheses are tested out to find out if any of them help resolve the problem.
3. ***C—Check:*** the results of the tests are analyzed.
4. ***A—Act:*** depending on the results, either the solution is implemented, modifications to the plan are made, or the idea is dropped. If action is taken, it is important to take steps to hold the gains that have been made. This is sometimes called "institutionalizing" the solution.

PDCA is a continuous cycle. After taking action, a new phase of the cycle begins and a new plan or set of hypotheses is developed. This framework helps the team or individual to stay on track and stay focused on the problem at hand. Repeated application of the PDCA cycle can lead to long-term improvements, and each new step leads to a higher level of quality.

The Plan-Do-Check-Act cycle

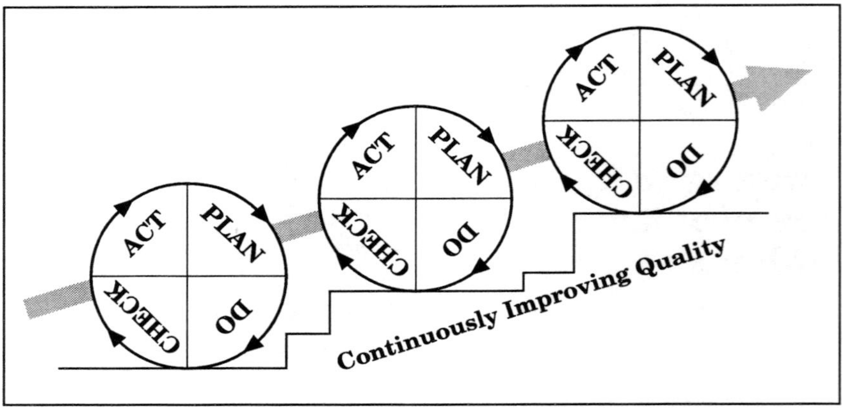

As we proceed through this chapter, an example of a very real printing situation will demonstrate the use of the PDCA cycle as well as showing how the problem-solving tool kit works under the framework of PDCA.

---

*Dr. Edwards Deming also referred to this as the *Plan-Do-Study-Act* process, in accordance with his theories about process improvement as a continuous process of learning through data collection and process study.

The Problem-Solving Tool Kit—An Introduction

As outlined in Chapter 1, a group of problem-solving tools has become known collectively as "The Seven Tools of Statistical Process Control." These tools were first grouped together as a cohesive set by the Japanese quality guru, Kaoru Ishikawa. Ishikawa continuously strove to find better ways to make team activities (quality circles in Japan) more effective. He saw these tools as being the most effective means for a team to accomplish its process improvement objectives. The tools can be utilized individually for specific purposes or as a group.

The seven tools of statistical process control are:

1. Flow charts
2. Cause-and-effect diagrams
3. Check sheets
4. Pareto diagrams
5. Scatter diagrams
6. Histograms
7. Control charts (and run charts)

As a group, the tools function as follows: At the beginning of a team's efforts, it first must define the process that it is attempting to improve. To this end, the process is flow-charted, identifying all of the steps, decisions, and critical checkpoints in the process. This is also known as "process mapping" as the team is actually developing a guideline to understanding the process. Next, the team identifies problem areas in the process and brainstorms about possible solutions to the various problem areas. This is done using the cause-and-effect diagram and various "thinking tools" such as the "why technique," "contingency method," and the "affinity diagram." The team must then collect data and analyze the data to determine the magnitude and nature of the problems or variations in the process. The data collection is facilitated using various check sheets and is analyzed using Pareto diagrams, scatter diagrams, histograms, and control charts. Run charts are also very useful for analyzing initial data. With a run chart, data points are simply plotted over time to find various types of patterns or trends. Finally, once an improvement has been made, the gains must be held. To ensure this, it is possible to use the control chart for detecting potential problem areas. Thus, the tools can be summarized by the phase of the problem solving where they apply, as shown in the table at the top of the next page.

Phases of Problem Solving and the Tools That Apply

Phase	Tool(s)
Process definition	Flow chart
Hypothesis generation	Cause-and-effect diagram
Data collection	Check sheets
Data analysis	Scatter diagram, Pareto analysis, histogram, control chart
Holding the gains	Control chart

The flow chart on page 91 shows how these tools are used together to address different aspects of continuous improvement over time.

Let us now turn our attention to the example that will be used throughout the rest of this chapter to demonstrate the applications of the seven tools. Remember that, at each stage of problem solving, we are in a phase of PDCA, which we will refer to in order to stay on track and to remain focused on our objective.

Our Example: Customer Satisfaction at Casey Press

Casey Press is a hypothetical, but typical, commercial printer founded in the 1950s and built from scratch to its present state. The company has annual sales of $7 million, primarily in the highly competitive four-color commercial market. The customers are mostly other small and medium-size businesses who use its services for their own advertising, promotional, and internal purposes.

Recently several complaints from customers have surfaced. These have included complaints about color consistency, poor folding on some brochures, and a couple of late deliveries. In the old days, the owner and president, Ms. Casey, would have thought of these as isolated incidents, but recently she has been doing some reading about quality management and wondered if the problem might be more systemic in nature. These readings have included the following information:

- Most customers don't complain; they just leave and you never find out why.
- It's not enough to satisfy customers; if you want them to be loyal, then they must be delighted.
- 85% of quality is the responsibility of management.
- If you want to improve quality, it's best to take a planned, scientific approach.

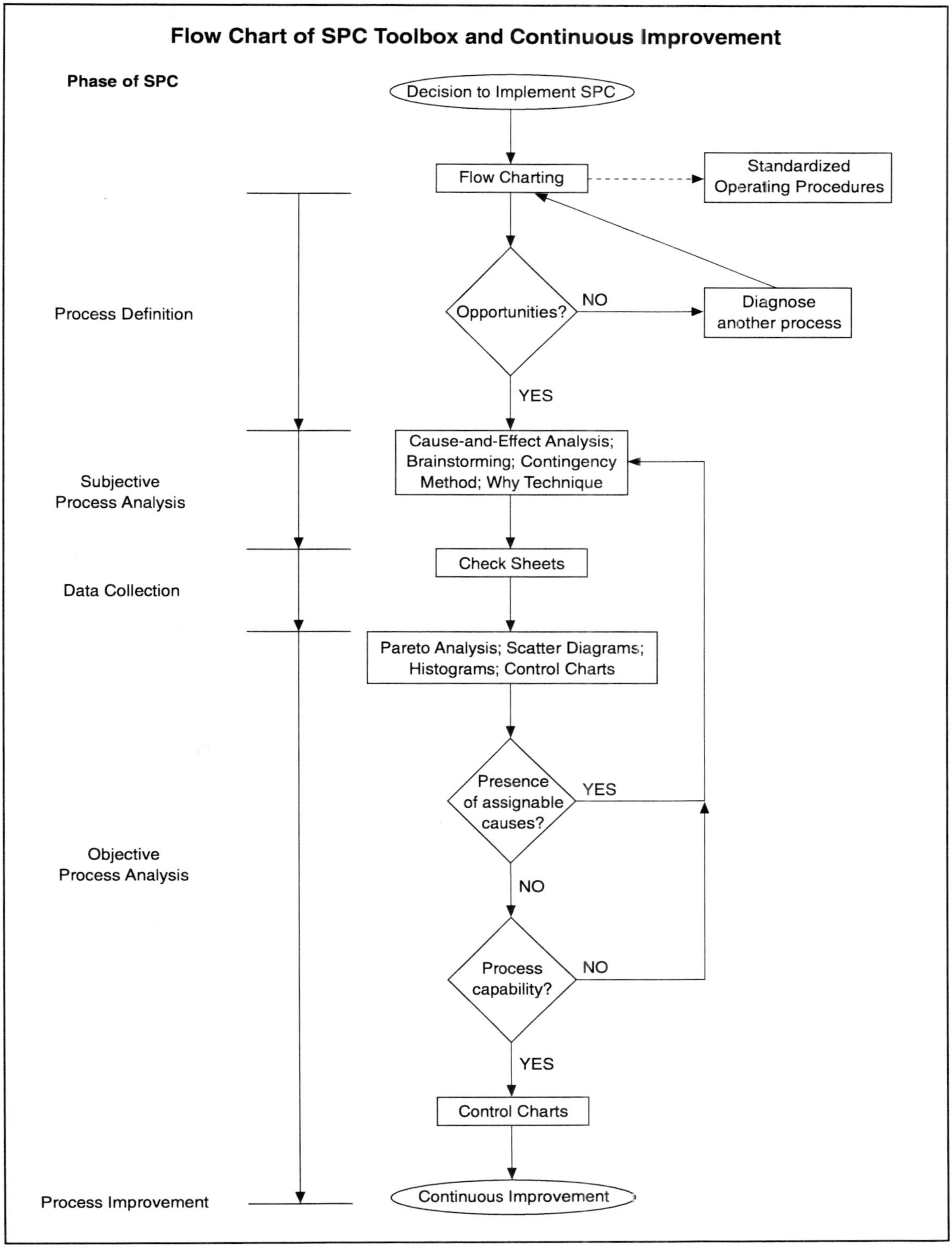
Flow Chart of SPC Toolbox and Continuous Improvement
Phase of SPC
Decision to Implement SPC
Flow Charting
Standardized Operating Procedures
Process Definition
Opportunities?
NO
Diagnose another process
YES
Cause-and-Effect Analysis; Brainstorming; Contingency Method; Why Technique
Subjective Process Analysis
Check Sheets
Data Collection
Pareto Analysis; Scatter Diagrams; Histograms; Control Charts
Presence of assignable causes?
YES
NO
Objective Process Analysis
Process capability?
NO
YES
Control Charts
Process Improvement
Continuous Improvement

- The best people to improve a process are usually the ones who actually do the work in that process.

Having read this, Casey decided to do a little investigating. The first thing she did was to conduct a customer survey of all of the customers who had come to the company in the past but not in the previous year. Asking why they hadn't returned in the past year, Casey found out the following from those who returned the survey:

- 10% had gone out of business.
- 25% had found somebody to do their printing cheaper.
- 15% had found somebody to do their printing more quickly.
- 30% had experienced quality problems with Casey Press and had therefore gone elsewhere.
- 20% had gone to other printers, but couldn't explain why.

Casey's first reaction to this information was typical: disbelief, having believed that Casey Press was price, schedule, and quality competitive. Apparently she had been mistaken, but how could she improve costs, on-time delivery, and quality at the same time. This seemed an impossible task until she came across something called the *Deming Chain Reaction*, which said that by improving and focusing on quality you would be able to reduce waste and in turn reduce costs, improve pricing flexibility, improve turnaround time, and therefore improve customer satisfaction.

Casey became convinced that this was the answer to her dilemma, so she sat down with the employees at the plant and began working on improving quality. Our tools will give you a picture of some of the things that were found, and done, to improve quality.

The Tools

Flow Charting the Process

The first thing to do when looking at improving a process is to actually define what the process is. To this end, a flow chart is the tool of choice. Flow charts represent a step-by-step breakdown of the particular process under investigation. They examine the actions and decisions that relate to each step in an operation and can give detailed information about potential sources of variation and inefficiency. They also give people a better feel for the big picture and where they fit into that picture.

Therefore, the definition of a flow chart is *a graphic description of the methodology involved in a given process*.

The following is a list of the symbols commonly used in creating a flow chart. While there are others, these are the ones most generally in use:

Symbol	Description
Start/finish	This symbol is used for defining the beginning and end (boundaries) of the process.
Activity	This symbol is used to show the activities utilized in the process (the actual steps).
Decision	This symbol shows when a decision is to be made and what the results will be.
Process flow →	This symbol connects the other symbols in sequence and shows how the process flows.

Flow charting is a logical way to organize your operation at the beginning of the SPC process. Since the goal of SPC is to improve the product by improving the process, it is necessary to understand the steps critical to the process at hand.

Flow charting helps people to identify routine items that might be neglected until something goes haywire. By creating flow charts, you create standardized operating procedures (SOPs). Thus, you are ensured of doing things correctly the first time, every time. Flow charts are very useful in identifying redundant or wasteful methods in an operation. A flow chart that identifies many loops and subsystems gives many opportunities to intervene and streamline the process. Flow charts help you to focus your process improvement efforts in specific, manageable areas.

Flow charting can serve as an excellent training exercise. By creating flow charts for their operations, employees gain the following:

- An understanding of the "big picture"
- An awareness of the importance of the many details of the operation that might otherwise be circumvented
- A feeling of fitting into the company as a whole, not just one specialized department
- Pride in knowing that their experience and opinions count

The best people to make the flow charts are the people actually involved in the activity in question. A flow chart should be developed for each and every one of your operations as well as major and critical procedures. A primary advantage of flow charts is their universal applicability to all areas of your operation, not just production but also sales, marketing, design, and finance.

Flow charting serves useful purposes both in the identification of key process variables and as a process unto itself.

Flow charting at Casey Press. At Casey Press, a cross-functional team of employees came up with a flow chart for the rough overall flow of the process as a job flows through from start to finish. This exercise proved beneficial, because for the first time all of the employees had to agree on the actual way in which they did things (not how they would have liked to do things). They also discovered many things that could be improved. Interestingly enough, the group noticed that not everyone did the same things the same way, even though they all had the same goal in mind. A rough version of the flow chart they came up with follows on page 95.

From this flow chart, the team members identified a number of key points that could help them in their quest to reduce costs, shorten turnaround times, and improve quality. For example, it seemed that the use of a final quality check would be too little too late—especially so since a problem found at that stage might necessitate starting over entirely from the beginning of production. Someone in the group came up with the idea that by performing internal quality control checks throughout the process, there might be the chance of avoiding problems downstream. One person suggested that each department should have its own quality control procedures. Casey thought this would be a good idea. Another person said that they needed to narrow their focus to the important junctures in the process. Finally, someone else suggested that they should collect some data to find out where the problems are occurring in order to find out how they might be solved. Everybody thought that this would be a good idea.

So they came up with a plan for collecting data, which was:

1. Brainstorm about the possible types of problems a job might have.
2. Develop a data collection check sheet to employ for a sample period of time.
3. Analyze the data to focus their efforts.

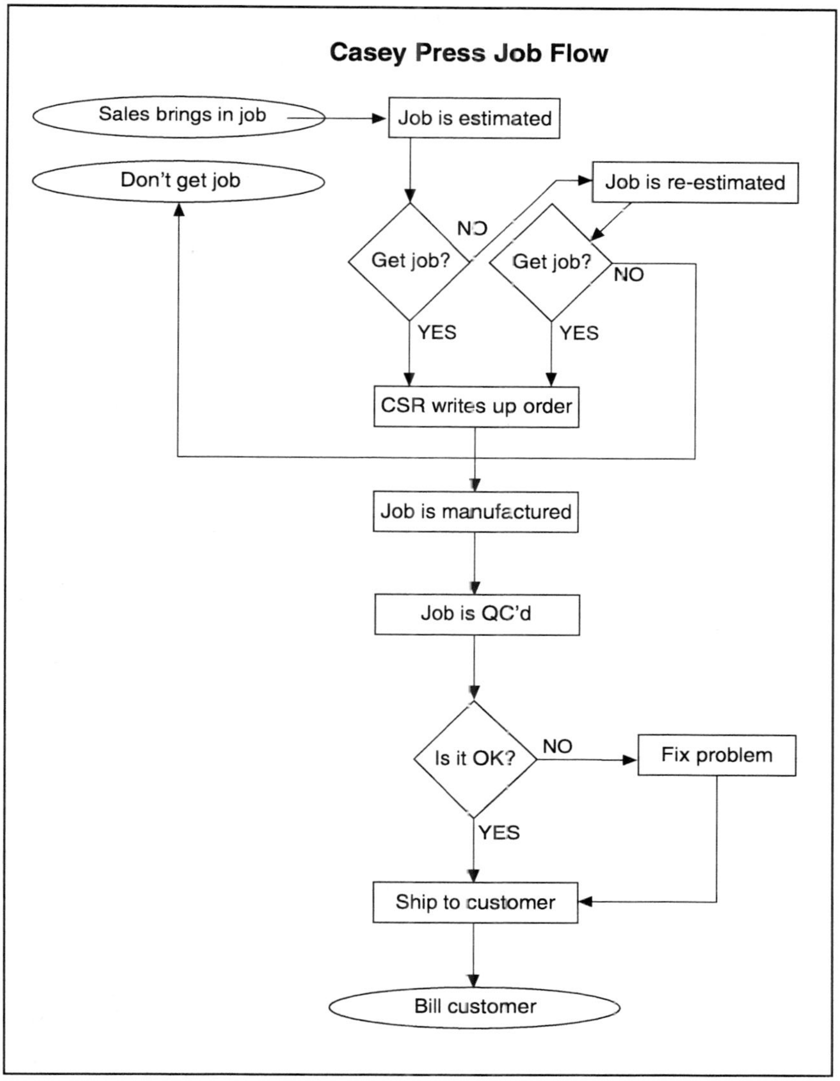

Brainstorming about Possible Solutions to Problems

The premise of brainstorming can be summed up in a single word: synergy. Synergy means simply that working as a group we can come up with a better, more effective solution than if we had worked on the problem as individuals. The reason for this is that we can bat ideas off of one another and stimulate each other's creative faculties. Several rules have evolved for brainstorming over time that help to make the process more effective. Also, several tools have evolved that allow us to organize the results of the brainstorming session into something meaningful as well as to make sure that the result of the brainstorming session is a solution which is both exhaustive as well as one which addresses the "root cause" of the problem, not just the symptom.

The rules for conducting a brainstorming session include the following:

1. State the problem clearly so that all members of the team agree as to what they are actually brainstorming about.
2. Write down the problem in a clear statement on a flip chart for all to see.
3. Give team members 5–10 minutes to individually write down their ideas about the problem (no discussion at this point, members work on their own).
4. Go around the group, in order, with each member presenting one idea when it is his or her turn. Each idea should be clearly expressed in as few words as possible and written down on a list on the flip chart.
5. At this point, there should be only one person talking and that is the person with the idea. The only interaction should be possibly between that person and the person writing the idea down, and this for clarification purposes only.
6. Go around the group, one idea per turn, until all ideas are exhausted.
7. If you don't have an idea when it is your turn to speak, just say "pass."
8. When all have said pass in succession, the brainstorming session is over and discussion can begin.

Here are a few other points to consider in this process:

- ***Never*** and under ***no circumstances*** should one member of the group criticize another's idea during the brainstorming session. It will inhibit the generation of ideas.
- Members should be constantly writing down new thoughts as other members present ideas that trigger one's own ideas.
- The goal is to synergize, so be creative. Remember that even the craziest idea might stimulate another member to come up with a great solution, and if that happens the process has been a success.

Note: An alternative to writing the ideas on a flip chart in a brainstorming session is for the facilitator to write the ideas on Post-It notes and then arrange these into groupings suggested by the team. This makes it possible to develop logical groupings of ideas and helps stimulate new ideas. This process is known as the "affinity diagram" method, as the team develops groups of ideas that have a natural *affinity* for each other.

Fishbone, or Cause-and-Effect, Analysis

The fishbone, or cause-and-effect, analysis is a useful tool for organizing the results of your brainstorming session. This type of analysis is good for helping to break down a factor into its many component parts. This will point you in the direction of where to implement SPC. The fishbone analysis takes into account the following "6 M's":

- **Man**—what is the effect of people in the process; how are they trained and motivated
- **Machine**—what role does equipment play in the process; is it suitable to do the task required or can it be improved
- **Milieu**—what is the environment like; this includes the physical environment (hot/cold, humid/dry, light/dark) as well as the cultural environment (friendly/hostile, trusting/deceitful, empowered/controlling)
- **Methods**—what are the procedures employed; are these documented as standard operating procedures; are they effective
- **Materials**—to what extent are the materials (paper, ink, rollers, blankets, film, chemistry) suited to the job; are the most cost-effective ones used or just the cheapest
- **Measurement**—how is the process measured and controlled (densitometry, waste and rework figures, on-time delivery, cost of quality, pH/conductivity, surveys of customer satisfaction)

Knowing that each of these can contribute to the problem, you can isolate the cause of that particular problem. The items listed in the fishbone diagram will give you all the data needed to analyze a particular problem and design experiments to verify or deny the causal relationship between an item and the factor it may have bearing upon. Remember that the goal of cause-and-effect analysis is to get a process into a stable and controlled state through the identification of major sources (causes) of variation.

Cause-and-effect analysis is a logical stepping stone between flow charting and data collection. Like flow charting, cause-and-effect analysis can be used in all areas of an operation, not just production. The tool enables you to identify problems to track back through the process to their origin.

It is generally the #1 method used in brainstorming sessions. Cause-and-effect analysis is a simple and powerful tool for getting people to work together in quality teams towards the elimination of quality problems. It is also called "fishbone" diagram (due to its appearance) or "Ishikawa" diagram (after its originator, Dr. Ishikawa). In constructing a cause-and-effect diagram, the following format is utilized. The causes are broken down on the left, and the effect or problem is shown on the right. The branches of the fishbone are causes that lead to one another.

General Format for a Cause-and-Effect Diagram

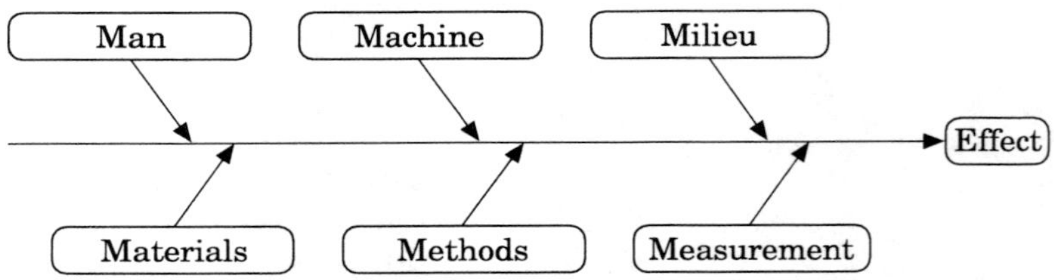

Two additional methods are worthy of mention here. These are the **contingency method** and the **why technique.** The contingency method is a fun way of looking at the issue from a different perspective. If your initial brainstorming session asks how to improve a process, then the contingency method looks at the opposite: come up with all of the ways to mess up or defeat the process. By doing so, it is possible to have a little fun and sometimes come up with an idea that otherwise might have been missed. This technique is very useful when people are hesitant to open up. The why technique is used to search deeper and deeper through the various layers of causation asking "why" at least five to seven times until you have discovered the root cause that can turn the problem on and off. For example, customers are unhappy with our late deliveries, WHY: because of scheduling difficulties, WHY: because the bindery is below productivity standard, WHY: because of excessive breakdowns, WHY: because no preventive maintenance, WHY: because it is not a high budget priority. Once you have identified the root cause (misplaced priorities as it relates to preventive maintenance, in this case), you can identify a solution that will serve the original need (to improve on-time delivery).

Brainstorming at Casey Press. You will recall that the employees at Casey Press were going to brainstorm about the many possible causes of job problems. This is exactly what they did. First, they stated the problem clearly and wrote the problem in a statement on the flip chart for all to read and see:

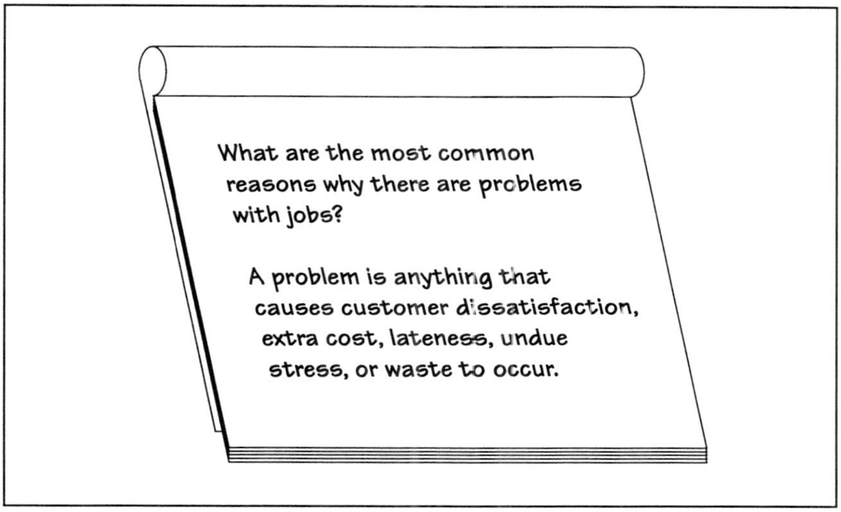

It should be noted here that this mandate is far too large for any team to solve. The next logical step after brainstorming about "problems with jobs" is to collect data and narrow down the scope of the team's effort. Next, each person was given a few minutes to think about the problem statement and write down as many of these as possible on a piece of paper on their own. After doing that, they began going around the group and getting one idea from each member at a time. As they did so, these were written on the flip chart.

Some of the items they came up with were:

- Lack of attention to detail
- Equipment in poor condition
- Lack of understanding of what the customer wants
- Press operators not trained
- Equipment not capable
- No measuring equipment for QC
- Use of poor materials (ink and paper)
- Hickeys
- Excessive makeready times on press
- Static electricity in prepress department

These were then organized into a fishbone diagram according to the six M's:

Man	• Lack of attention to detail • Lack of understanding of what the customer wants
Machine	• Equipment in poor condition • Equipment not capable
Material	• Use of poor materials (ink and paper)
Milieu	• Hickeys • Static electricity in prepress department
Measurement	• No measuring equipment for QC
Methods	• Press operators not trained • Excessive makeready times on press

Remember that these are only a few of the ideas the group came up with. In reality, a group might come up with fifty or more items in a brainstorming session like this. If the group comes up with a huge list, the group can use the **multivoting technique** to narrow the list down to a more manageable number. In multivoting, each member of the group gets a certain number of votes. The square root of the number of items is usually a good number of votes for each member to have. Therefore, if the team came up with fifty items, then each member would get seven votes (square root of 50 rounded off). These votes are then cast according to the preferences of each individual. For example, one person might feel strongly that a single item is the most important and give that item four votes. The other three votes might go to two other items, one item getting two votes and another item getting one vote. The key is that individuals allocate their votes according to their priorities and unique outlook. After voting, the team narrows down the list to the top items (usually no more than fifteen) and investigates these further.

The team also used the contingency method as part of its brainstorming session. In using the contingency method, it asked the question: "What can we do to cause customer dissatisfaction?" In response to this question, it came up with several new items for its list. These were:

- Print the wrong PMS color
- Charge the wrong amount for the job
- Deliver the job later than promised
- Deliver the job to the billing (not shipping) address, if different

- Print the job out of register
- Plugging in the shadows on critical halftones
- Color variation not controlled

Now that the team had developed a list of likely causes of quality problems and customer dissatisfaction, it was ready to go into the mode of collecting data so that the issue could then be analyzed with less subjectivity. To do this, the team developed a checklist.

Using a Check Sheet or Checklist to Collect Data

Check sheets are graphic data collection tools that allow the user to simultaneously collect, compile, and interpret information. A well-designed check sheet makes data collection easy and data analysis automatic. There are essentially five formats for the check sheet, each of which can be varied and tailored to the specific circumstances at hand.

The five types of check sheets are:

1. Production process distribution check sheet
2. Defective item check sheet
3. Defect location check sheet
4. Defect cause check sheet
5. Confirmation/inspection checklist

Pages 104–112 show examples of the five types of check sheets.

A **production process distribution check sheet** is used for analyzing the frequency distribution of continuous data. Examples where this type of check sheet could be applied include printing characteristic variation, folder variation, gluing variation, and trimming variation. The result of using this type of check sheet is a histogram, which can be analyzed for process centering and spread. The two primary questions that the user of this check sheet must answer are:

1. What is the shape of the distribution (bell shaped, centering, spread, skew...)?
2. How does the distribution compare to process specifications?

Figures **1A** and **1B** (pages 104–105) show a production process distribution check sheet for diecut variation, in which the distance between the cut and the print has been measured. In this example, the process is shown to be fully capable of meeting requirements.

A **defective item check sheet** is used to collect data on the frequency of occurrence for various types of defects. This breakdown allows the user to categorize and prioritize defects for further analysis of causes. This information is often converted into a Pareto analysis for action on the vital few quality problems. Additionally, this information can be useful in cost-benefit analysis for specific quality improvement projects or for comparing the before and after situations when quality improvements are instituted. Figures **2A** and **2B** (pages 106–107) demonstrate the use of a defective item check sheet. In this example, various types of printing defects are examined. A defect is an item that fails to meet a specified quality standard. A defective item may contain one or more defects.

A **defect location check sheet** is a useful tool for appearance-related defects and is thus especially useful in the printing industry. The defect location check sheet consists of a diagram of the object under consideration, on which notations are made in the places where defects occur. This diagram helps the printer to isolate and pinpoint sources of printing defects. Figures **3A** and **3B** (pages 108–109) demonstrate the use of a defect location check sheet. In this example, a form with nine-up folding cartons is examined for printing defects. The diagram shows that hickeys, scumming, and plugging are all occurring in varying frequencies. Furthermore, by analyzing the location of the defects, it is possible to troubleshoot causes. For example, the hickeys are virtually all on one side of the form, which might indicate a source of dust and dirt, such as a light fixture, on that side of the press.

A **defect cause check sheet** is similar to the defective item check sheet, except that instead of simply listing defects, it attempts to link causes to effects by stratifying along lines such as by machine, by shift, by day, and by operator. Figures **4A** and **4B** (pages 110–111) show a defect cause check sheet that examines finishing defects. Analysis of this check sheet shows where the greatest frequency of trim, staple, and collating problems can be traced to in terms of shift, day, operator, and machine. This information can help to pinpoint key causes very quickly.

An **inspection checklist** is simply a questionnaire designed to ensure that critical parameters aren't overlooked. (See Figure **5** on page 112.) It is simply a list of "OKs" that must be checked in order for the process to continue. Each point on the check sheet is roughly analogous to the decision juncture on the process flow diagram. Checklists are ideal instruments for use on incoming copy, shipping and receiving, platemaking operations, preflighting incoming customer computer disks, and any other place where critical quality parameters can be defined by a simple “yes” or “no.” Interestingly, inspection checklists are one of the easiest ways of instituting effective record keeping procedures to support the ISO 9000 process. Each checklist can be dated and signed and then becomes a quality record that an in-process inspection has taken place. This makes internal auditing of a system very easy to institute.

Developing a useful check sheet is not an easy task. It requires input from a variety of sources, including internal and external suppliers and customers, and operational personnel at the source of the defects being examined. Furthermore, it is necessary that people realize the true purpose of the check sheet and don’t believe that this is a mechanism for scrutinizing their performance. Finally, in order to develop a truly useful check sheet, it is important that one first master the skills of brainstorming, cause-and-effect analysis, flow charting, Pareto analysis, and basic statistics, each of which can contribute to the final check sheet. Brainstorming and cause-and-effect analysis are particularly useful when deciding which attributes should be checked and what their possible causes are. Flow charting is useful in deciding where the best strategic location in the process is to place the check sheet and what the impact is of a detected defect. Pareto analysis can be applied for the further analysis of information collected in the check sheet, especially the defective item type. Lastly, basic statistics is necessary in order to correctly interpret and use the production process distribution check sheet.

If developed and used correctly, the five types of check sheets can be extremely valuable tools in the quality improvement tool kit. Each has a particular strength in terms of what type of situation it applies to; however, all five have in common their ability to allow the operator to easily collect and interpret data in a fashion that is meaningful to their job. Before implementing or even developing check sheets, however, it is necessary that people develop other skills, such as brainstorming, so that the check sheet will ultimately represent a usable instrument. Keep in mind, too, that check sheets often go through multiple revisions in their early stages until a sound data collection system is in place.

Figure 1A

Casey Press
Production Process Distribution Check Sheet

Plant ____

Date ____

Product ____

Machine ____

Operator ____

Measurement ____

Calibration tool ____

Remarks ____

____

____

Dimension																		
Count 20																		
19																		
18																		
17																		
16																		
15																		
14																		
13																		
12																		
11																		
10																		
9																		
8																		
7																		
6																		
5																		
4																		
3																		
2																		
1																		
Total Frequency																		

Figure 1B

Casey Press
Production Process Distribution Check Sheet

Plant: Main
Date: April 1995
Product: folding cartons: print-to-cut dimensions
Machine: Bobst
Operator: group
Measurement: Mitutoyo 50x scope
Calibration tool: yes—before experiment
Remarks: experiment to determine process capability at ±1/64 in., see graph for results

Dimension	.485	.486	.487	.488	.489	.490	.491	.492	.493	.494	.495	.496	.497	.498	.499	.500	.501	.502
Count																		
20																		
19																		
18																		
17																		
16																		
15																		
14											x							
13							x				x							
12							x				x							
11							x		x	x	x							
10					x		x		x	x	x							
9					x		x	x	x	x	x							
8					x		x	x	x	x	x							
7					x	x	x	x	x	x	x	x						
6				x	x	x	x	x	x	x	x	x						
5				x	x	x	x	x	x	x	x	x						
4				x	x	x	x	x	x	x	x	x						
3				x	x	x	x	x	x	x	x	x						
2	x			x	x	x	x	x	x	x	x	x			x			
1	x	x	x	x	x	x	x	x	x	x	x	x	x	x	x	x	x	
Total Frequency	2	1	1	6	10	7	13	9	11	11	14	7	1	1	2	1	1	

Figure 2A

Casey Press
Defective Item Check Sheet

Plant ______________________________

Date ______________________________

Product ______________________________

Machine ______________________________

Operator ______________________________

Characteristics ______________________________

Remarks ______________________________

____

____

Defect Type	Check	Subtotal
	Grand Total:	
Total Defectives		

Figure 2B

Casey Press
Defective Item Check Sheet

Plant Main

Date April 1995

Product covers

Machine N/A

Operator N/A

Characteristics critical defects

Remarks analysis shows hickeys, folder marks, and glue as high-impact items

Defect Type	Check	Subtotal
Typographical	xxx	3
Imposition	xxxx	4
Scumming	xx	2
Ghosting	xxxxx xx	7
Hickeys	xxxxx xxxxx xxxxx xxx	18
Folder Marks	xxxxx xxxxx xxxx	14
Setoff	x	1
Wrinkling	xxxxx xxx	8
Trim	xxxxx	5
Staples	–	0
Glue	xxxxx xxxxx xxxxx xxxx	19
Others	xx	2
	Grand Total:	83
Total Defectives	xxxxx xxxxx xxxxx xxxxx xxxx	24

Figure 3A

Casey Press
Defective Location Check Sheet

Plant ________________________________

Date ________________________________

Product ______________________________

Machine ______________________________

Operator _____________________________

Characteristics ________________________

Remarks ______________________________

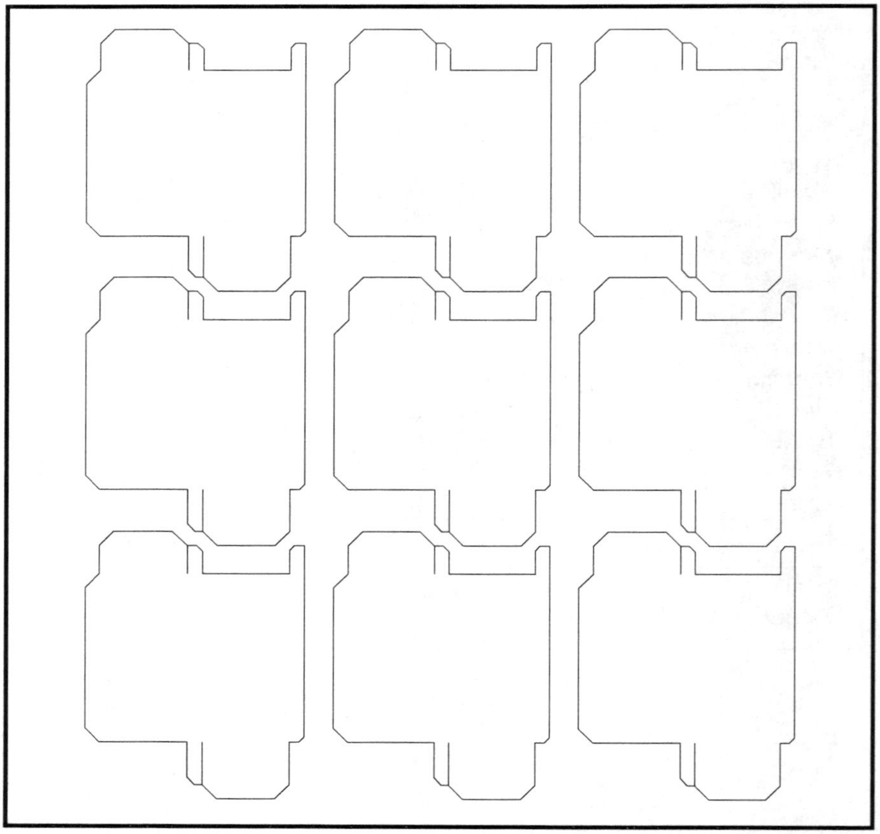

Figure 3B

Casey Press
Defective Location Check Sheet

Plant: Main
Date: April 3–8, 1995
Product: nine-up carton
Machine: Speedmaster
Operator: Mike
Characteristics: critical defects
Remarks: diagram shows roller problems, TIR, blanket swelling, hickey factory

Legend
x = hickeys
s = scumming
○ = plugging

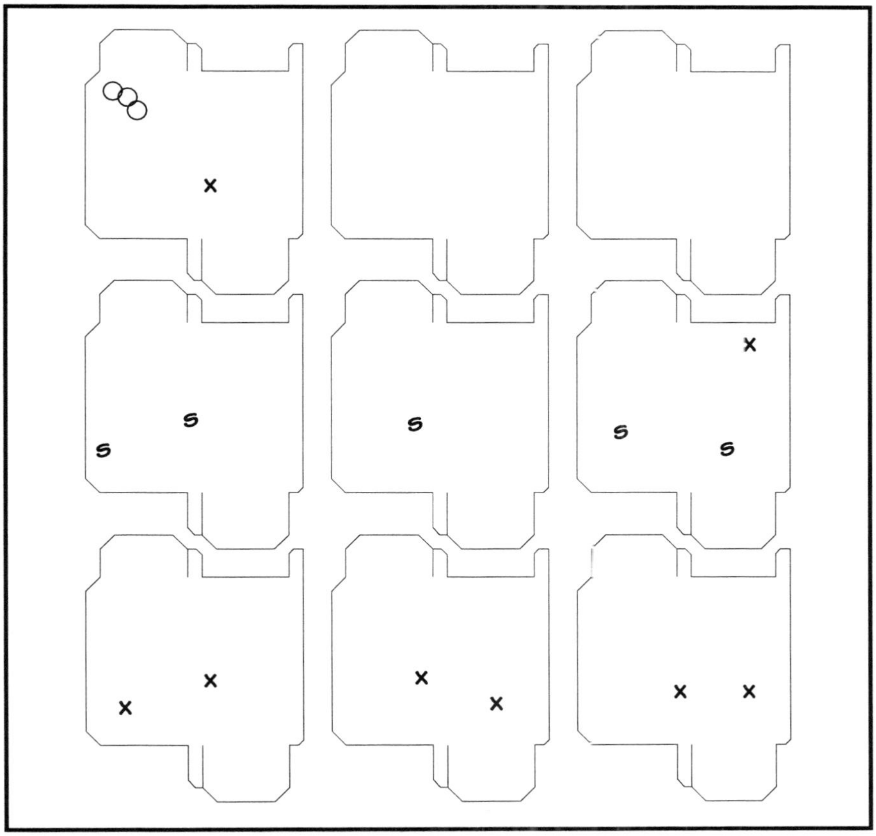

running direction →

Figure 4A

Casey Press
Defect Cause Check Sheet

Plant ______________________________

Date ______________________________

Product ______________________________

Machine ______________________________

Operator ______________________________

Characteristics ______________________________

Remarks ______________________________

____

____

Legend:

Figure 4B

Casey Press
Defect Cause Check Sheet

Plant Main
Date April 11–16, 1995
Product booklet
Machine Muller Martini
Operator Liz, Hank, Kersten, Mike
Characteristics finishing defects
Remarks analysis compares operators, shifts, and machines

		Monday		Tuesday		Wednesday		Thursday		Friday	
Equipment	Worker	am	pm	am	pm	am	pm	am	pm	am	pm
Machine 1	A	oox	ox	ooo	oxx	ooox xx	oooo xxx	oooo x	oxx	oooo	oo
	B	oxx	ooox xo	oooo ooxx	ooox x	oooo ooxx	oooo oox	oooo ox	ooox	ooxx	oooo o
Machine 2	C	oox	ox	oo		oooo o	oooo oox	oo	o	oot	oo
	D	oox	ox	oo	ooo	ooo t	oooo ox	oo	oot	oott	o

Legend:
o = missing staples
x = miscollated
t = trim alignment

Figure 5

Casey Press
Camera-Ready Art Checklist

Person Checking ______________________ Date __________

Job Name ______________________ Job Number ______

____

____

	OK	Fix*	Date Fixed	Allowable Exception(s)**
Condition of Material				
Materials used for pasteup board	☐	☐	________	☐
Materials used for overlays	☐	☐	________	☐
Adhesion of text	☐	☐	________	☐
Adhesion of art	☐	☐	________	☐
Position of register and trim marks	☐	☐	________	☐
Type resolution	☐	☐	________	☐
Type density	☐	☐	________	☐
Layout				
Overall dimensions	☐	☐	________	☐
All elements are in position	☐	☐	________	☐
All elements are aligned	☐	☐	________	☐
All illustrations are present	☐	☐	________	☐

*Deviation(s) to be fixed ______________________________

____

____

**Exception(s) allowed ______________________________

____

____

The data collection check sheet at Casey Press. The employees at Casey Press took the items that they had come up with in their brainstorming session and developed a check sheet from this. Each department had a copy of the check sheet, and each time one of the listed items was observed by that department, it was noted on the check sheet by indicating the job number where this happened. In that way, it would be possible to determine if the incident had already been detected by another department as well as to identify if an incident occurred more than once for a given job. The checklist that the group came up with is shown on page 114.

The group decided to use this tool for one month to see what the results would be. Each department would fill out its own check sheet. These would then be compiled by the company controller using his spreadsheet program. The results would be analyzed using a Pareto analysis.

Ms. Casey emphasized, before they started collecting data, that the information was critical to making improvements, which would allow them to increase productivity, quality, and customer satisfaction. Furthermore, she emphasized that the data would need to be accurate in order to make sound decisions. Finally, she stressed that in no way would the data be used **against** anybody or any department. She felt it important to state this final point so that people wouldn't be on the defensive, especially since the changes they were trying to make were positive and would benefit everybody.

Pareto Analysis

Pareto analysis is based on the principle that 80% of the effects in a given area (let's say paper waste) is caused by 20% of the contributing factors in that area. By eliminating just one big factor, it may be possible to eliminate as much of your problem as if you were to go after many smaller factors. Unfortunately, the smaller factors may, and often do, cloud the issue until the problem is studied and a Pareto Analysis is performed. As an example, let us take paper waste in the press department. First, try and isolate all of the possible causes of paper waste. These could include top and core waste, splice waste, makeready waste, and waste from product rejection. Next, monitor the amount of waste resulting from each of these and rank them as a percentage of total paper waste from most to least. By attacking the largest single cause, you will cover much more ground than if you had attacked other smaller causes.

Casey Press
Data Collection Check Sheet

Directions: If there is a production problem or customer complaint with a job, please indicate what the problem was (put the job number in the column) and the severity of the problem (1: minor incident; 2: major incident; 3: critical problem).

Department ____________________ **Date** __________ **Sheet #** ____________

Problem Type	Minor	Major	Critical	Total	%
Lack of attention to detail					
Equipment in poor condition					
Lack of understanding of what customer wants					
Press operators not trained					
Equipment not capable					
No measuring equipment for QC					
Use of poor materials (ink and paper)					
Hickeys					
Excessive makeready times on press					
Static electricity in prepress					
Print the wrong PMS color					
Charge the wrong amount for the job					
Deliver the job later than promised					
Deliver the job to the billing (not shipping) address, if different					
Print the job out of register					
Plugging in the shadows on critical halftones					
Color variation not controlled					
Other					
	Total	Total	Total	Total	
	%	%	%	%	

The Pareto Principle (80% of the effects are due to only 20% of the causes) is also known as the "80-20 Rule." What it means in quality control is that there are always a "vital few" causes that must be isolated from the "trivial many" other causes. Once the "vital few" are uncovered, they must be isolated and eliminated.

Pareto's Law was originally developed by the 19th century Italian economist Vilfredo Pareto, who observed that 80% of his country's wealth was owned by only 20% of its citizens. Dr. Joseph Juran applied this idea to quality and formulated the concept of the "vital few" problems over the "trivial many." If you were to attack the trivial many, which is often the case since they are often the most obvious, you are defeating the purpose since the cure is more costly than the potential gain.

A Pareto diagram is a bar graph used to show the relative importance of a given set of causes. Typical areas of concern that lend themselves to Pareto diagramming are quality costs, causes of waste, causes of variation, causes of customer dissatisfaction, causes of job reruns, and incoming materials variation. Pareto analysis is so simple that it is often underrated as a key tool in the quality improvement process. By gathering data and presenting it in a format that isolates vital areas from trivial areas, Pareto analysis helps to achieve agreement and consensus on which direction to take in the SPC process. The following methodology can be used for conducting a Pareto analysis:

1. Using cause-and-effect analysis, determine the area of concern (effect) and all of the known probable areas (causes) that can contribute to the effect.
2. Isolate the 5–10 most likely problem cause areas.
3. Determine a method to quantify each cause area.
4. Select an appropriate time frame in which to gather data on each cause area. The time frame should be based on the minimum amount of time required to get a representative, random sampling.
5. Gather data.
6. Graph data. This is done by creating bars in descending order from left to right according to quantity. The quantity corresponding to the height of the bars is listed on the left vertical axis. On the right vertical axis, list the cumulative percentages for the data from 0–100%. Draw a line graph from left to right for the cumulative percentages.
7. Go to work on the "vital few."

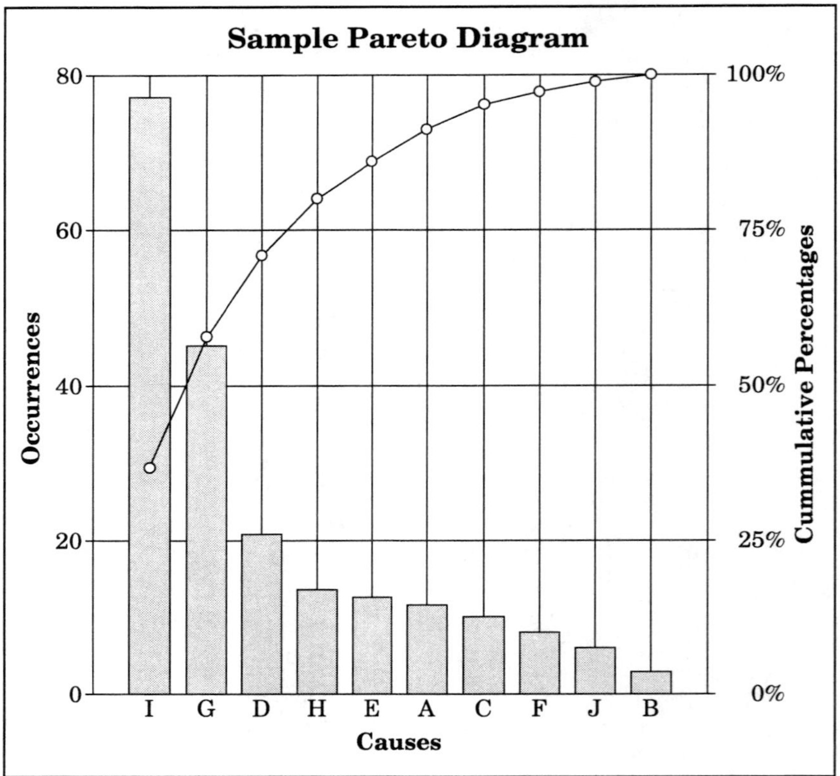

Note: Sometimes there are no "vital few." In these cases, the Pareto graph will be relatively flat. What this indicates is the lack of "special" or "assignable" causes of variation. The variation in the process is more or less common or systemic. Process improvement in such cases means addressing the entire system rather than a special cause. Such cases are rare in the early stages of SPC, as processes tend to be out of control at this stage.

Pareto analysis at Casey Press. The reader will recall that at Casey Press the team had developed a data collection check sheet to quantify its problems and would then conduct a Pareto analysis of this data in order to focus its efforts. After one month of using the tool, the results were summarized as shown in the first table on page 117 (for areas identified as critical problems).

The second table on page 117 shows the necessary formatting and calculations for putting together the Pareto diagram. The table has been sorted so that the items go from highest to lowest in terms of the number of jobs affected, except for the

Summary of Data Collection at Casey Press

Item	Number of jobs affected
Lack of attention to detail	7
Equipment in poor condition	13
Lack of understanding of what customer wants	4
Press operators not trained	2
Equipment not capable	9
No measuring equipment for QC	5
Use of poor materials (ink and paper)	3
Hickeys	2
Excessive makeready times on press	4
Static electricity in prepress	12
Print the wrong PMS color	1
Charge the wrong amount for the job	0
Deliver the job later than promised	2
Deliver the job to the billing (not shipping) address, if different	1
Print the job out of register	2
Plugging in the shadows on critical halftones	0
Color variation not controlled	2
Other (items not originally considered)	5

Formatting and Calculations to Develop a Pareto Diagram at Casey Press

Item	# of jobs affected	Percent	Cumulative percentages
Equipment in poor condition	13	17.6%	17.6%
Static electricity in prepress	12	16.2	33.8
Equipment not capable	9	12.2	45.9
Lack of attention to detail	7	9.5	55.4
No measuring equipment for QC	5	6.8	62.2
Lack of understanding of what customer wants	4	5.4	67.6
Excessive makeready times on press	4	5.4	73.0
Use of poor materials (ink and paper)	3	4.1	77.0
Press operators not trained	2	2.7	79.7
Hickeys	2	2.7	82.4
Deliver the job later than promised	2	2.7	85.1
Print the job out of register	2	2.7	87.8
Color variation not controlled	2	2.7	90.5
Print the wrong PMS color	1	1.4	91.9
Deliver the job to the billing (not shipping) address, if different	1	1.4	93.2
Charge the wrong amount for the job	0	0.0	93.2
Plugging in the shadows on critical halftones	0	0.0	93.2
Other (items not originally considered)	5	6.8	100.0

"other" category which is left, by convention, last. In this case, "other" is left as a group until an item within this group becomes larger than the next largest category. This item is then taken out of the "other" group and becomes a new category. For each item, we have calculated the percentage of jobs affected by that item in the third column. The fourth column is simply the cumulative percentage created by adding sequentially each of the percentages shown in the third column.

When put into the form of a Pareto diagram, the Casey Press data appears as shown in the illustration on page 119.

From this analysis, the team decided to focus in on the largest four areas as a means for reducing problems with jobs. These four areas were equipment condition, static in prepress, equipment capability, and lack of attention to detail. It is interesting to note that three of the four largest contributors to the problems were system-related and only one of the four was directly attributable to operators within the system. This, of course, is in line with Dr. Deming's 85:15 rule, which states that 85% of quality is a function of management, with only 15% being operator controllable. Intuitively, we would probably all agree with this rule, especially when one considers that management is responsible for such things as equipment acquisition and maintenance, training programs, purchase of materials and supplies, and development of standard operating procedures. The other thing that is interesting about the analysis at Casey Press is that there are no real easy solutions to the fundamental problems—so don't waste time looking for quick fixes. If you want real improvements in quality and customer satisfaction, you have to be willing to make major changes to the way you are presently doing business.

Scatter Diagrams

Evaluating and quantifying cause-and-effect relationships with the scatter diagram. A less common but often indispensable tool in the quality improvement toolbox is the scatter diagram, also known as a scattergram or correlation diagram. A scattergram is a tool used to determine the strength and nature of the relationship between two different variables. In other words, we can use a scattergram as a measurement tool for investigating cause-and-effect relationships in simple experiments.

When using scatter diagrams, the variable causing the problem is called the "independent variable" and is plotted on the x-axis of the graph. The variable being affected is called the "dependent variable" and is plotted on the y-axis of the

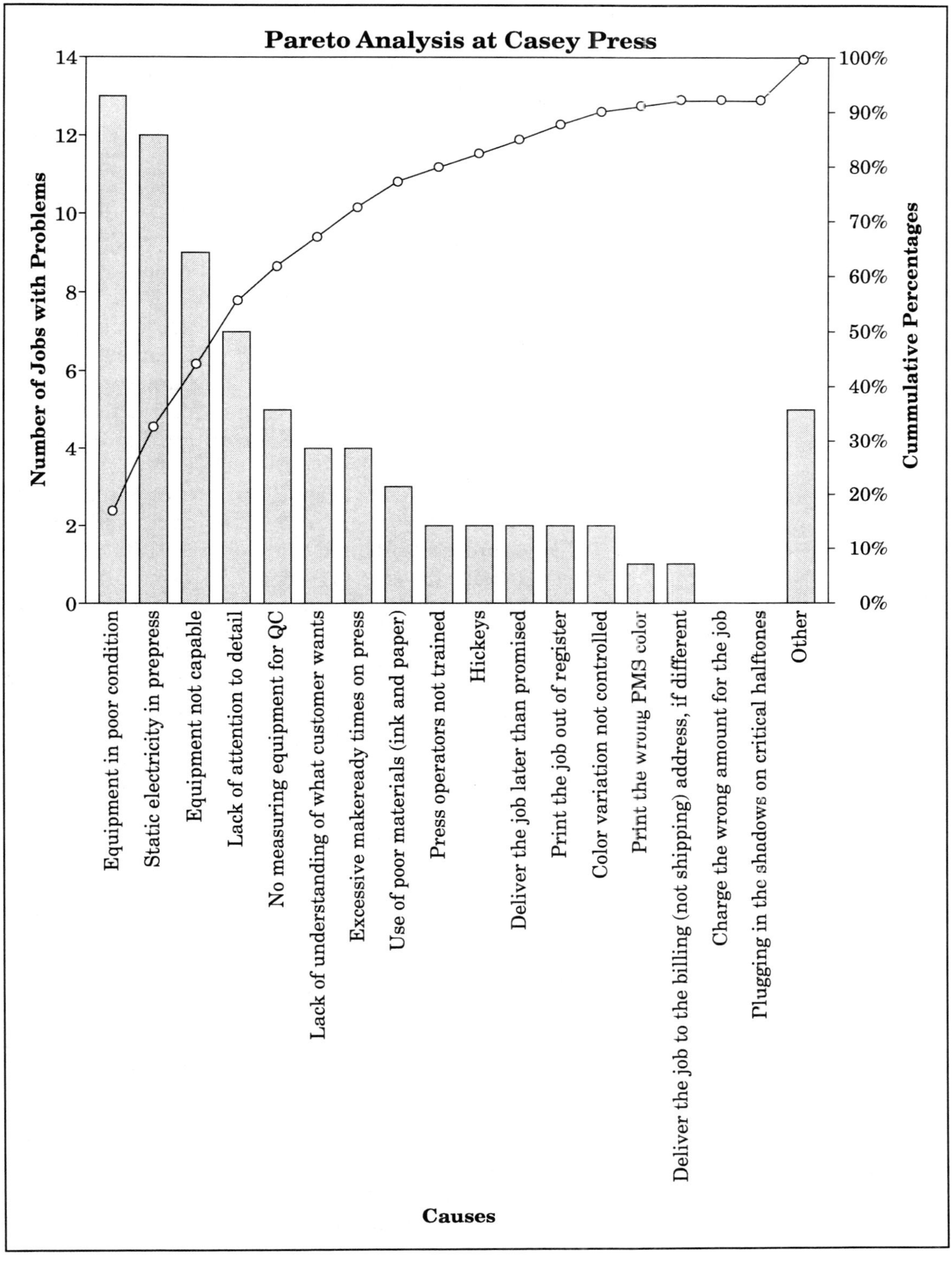

Pareto Analysis at Casey Press
Number of Jobs with Problems
Cummulative Percentages
0
2
4
6
8
10
12
14
0%
10%
20%
30%
40%
50%
60%
70%
80%
90%
100%
Equipment in poor condition
Static electricity in prepress
Equipment not capable
Lack of attention to detail
No measuring equipment for QC
Lack of understanding of what customer wants
Excessive makeready times on press
Use of poor materials (ink and paper)
Press operators not trained
Hickeys
Deliver the job later than promised
Print the job out of register
Color variation not controlled
Print the wrong PMS color
Deliver the job to the billing (not shipping) address, if different
Charge the wrong amount for the job
Plugging in the shadows on critical halftones
Other
Causes

graph. Our procedure is to systematically change the independent variable and measure the results. If there is a corresponding change in the dependent variable, we know that there is a relationship between the two variables.

If the two variables are closely related, they are said to be highly "correlated." For example, there is a high correlation between the skill level of a press operator and the quality of work produced. In this case, "skill" would be the independent variable, and "level of quality" would be the dependent variable. Let us look at this as an example. It is possible to investigate this relationship by taking a survey of similar printing companies. At each company, you would ask two questions: (1) "How many hours of training do your press operators receive per year?" and (2) "What is your average percent defective product?" Question #1 is a means for quantifying the independent variable "skill," and question #2 is a means for quantifying the dependent variable "level of quality." Let us suppose that the theoretical results shown in the table on page 121 were obtained from a survey of printing companies whereby 30 companies responded (30 is a generally accepted minimum sample size for this type of study; fewer data points might lead you to erroneous conclusions). As a scatter diagram, this data would appear as shown below.

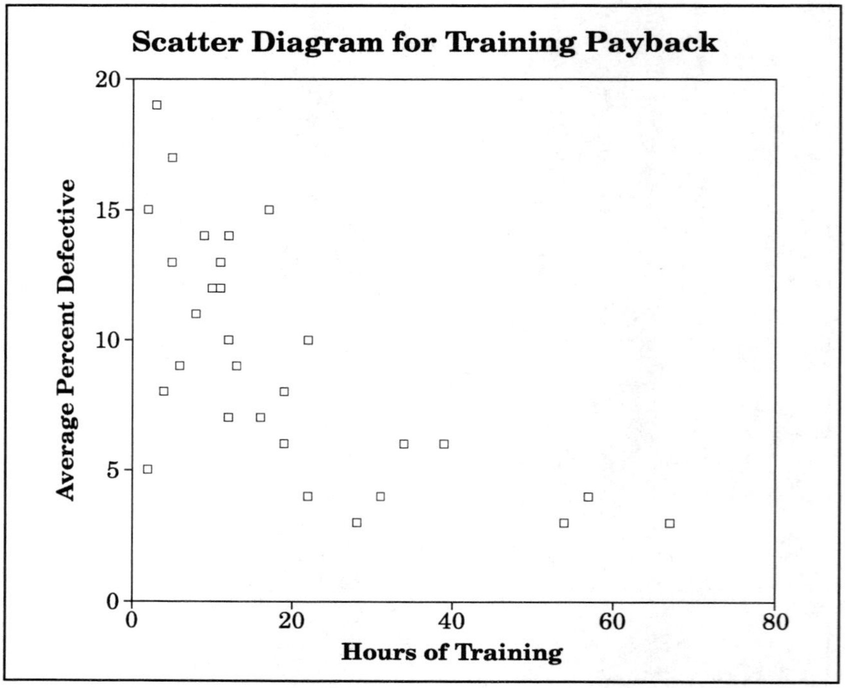

Company	Hours of training per year	Average % defective product
1	5	17
2	12	14
3	3	19
4	54	3
5	34	6
6	10	12
7	2	15
8	13	9
9	57	4
10	9	14
11	11	13
12	17	15
13	22	10
14	8	11
15	39	6
16	12	10
17	5	13
18	2	5
19	19	8
20	16	7
21	12	7
22	67	3
23	11	12
24	22	4
25	6	9
26	4	8
27	19	6
28	31	4
29	16	7
30	28	3

From this diagram, we can see that there is, in fact, a pattern to the data. As training hours increase, there is a tendency for the percent defective product to decrease. Obviously, this is not a perfect relationship, as there are many more factors affecting quality than just training; however, this does appear to be an important relationship. If this was a perfect relationship, then all of the points would fall upon a straight line. The more scattered the points, the more difficult it will be to identify a pattern, thus indicating the apparent lack of a significant causal relationship. Coincidentally, on the chart each dot represents what is known as a "paired sample." A paired sample is the description of an item in terms of its independent and dependent variables. For example, company 1's paired sample is 5 hours and 17% defective product. Where 5 hours and 17% intersect on the graph, we have the dot representing company 1. Thus, each dot is a different company as described by the two variables being investigated.

Scattergrams are often used to further investigate the strength and nature of relationships that have been identified by cause-and-effect analysis and Pareto analysis.

When a scattergram indicates a strong relationship, it is possible to fit a line through the trend. This is sometimes done for the following reasons:

- Prediction and forecasting of process events: it is possible to identify what will happen given a certain level of input.
- Optimizing your operating conditions: it is possible to determine what would be the best input to optimize a desired level of output.

Methodology. There are two different ways of fitting a line through a scattergram:

1. By eye
2. By mathematical regression analysis

Each of these methods can yield suitable results in terms of generating a line that helps you to interpret the data. The one big advantage of the regression analysis is that it gives you a number which indicates the strength of the relationship as well.

In order to fit a line through a scatter diagram **by eye,** you should apply the following steps:

1. Collect paired sample data. Plot the expected cause (independent variable) along the x-axis and the expected effect (dependent variable) on the y-axis. Plot the desired points on graph paper.
2. Decide if the scattergram shows a relationship worthy of further investigation. This is a matter of subjectively judging whether there is in fact a discernible pattern.
3. Use a straight-edge to estimate a line that goes down the middle of the plotted points in the direction of the trend that these points create.
4. Count the points on either side of the line. If they are approximately equal, the line is probably okay.

5. Estimate plus-or-minus boundaries by drawing two lines (one on each side of the center line) parallel to the center line that are *at equal distances from the center line and encompass all points.*
6. Make whatever predictions you desire using the center line as the nominal value and the outer lines as plus-and-minus values.

It is important to identify what are known as ***outliers*** in any analysis of scattergram data. An outlier is an obvious exception to the trend, if there is one. If there is an outlier, you will want to determine the explanation for its existence. If you can discover an explanation, this might help to indicate another possible cause-and-effect relationship. Suppose, for example, that the team discovered a company which invested nothing in training time, yet had only 2% defective product. This might indicate other possibilities for improving quality, such as better application of new technology or strong supplier partnerships to ensure high-quality supplies.

The following examples indicate the patterns that might be found with a scatter diagram as well as the steps to the eyeball method of fitting a line to the scatter diagram:

Possible scatter diagram patterns

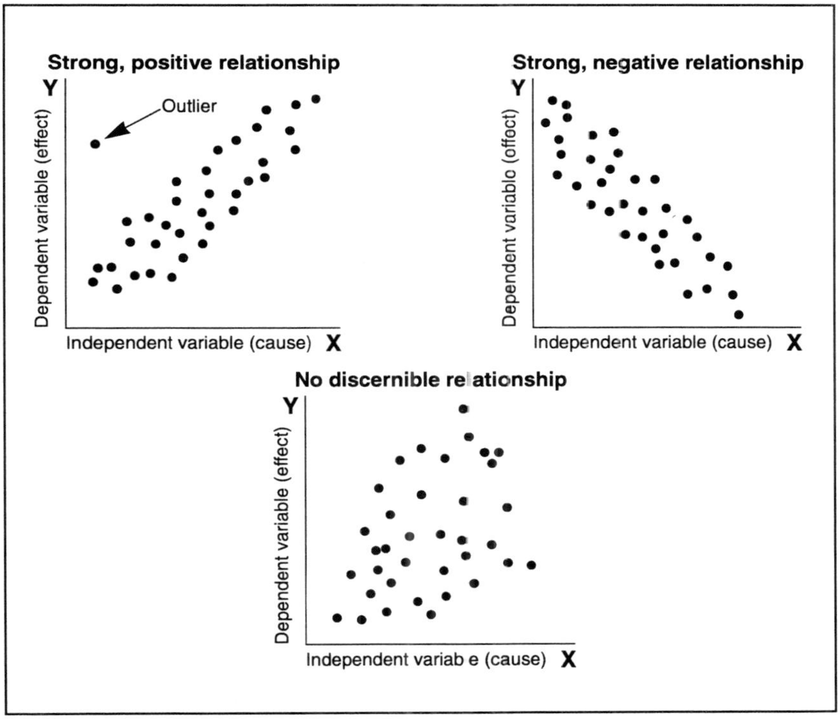

Using the eyeball method of fitting a line to the scatter diagram

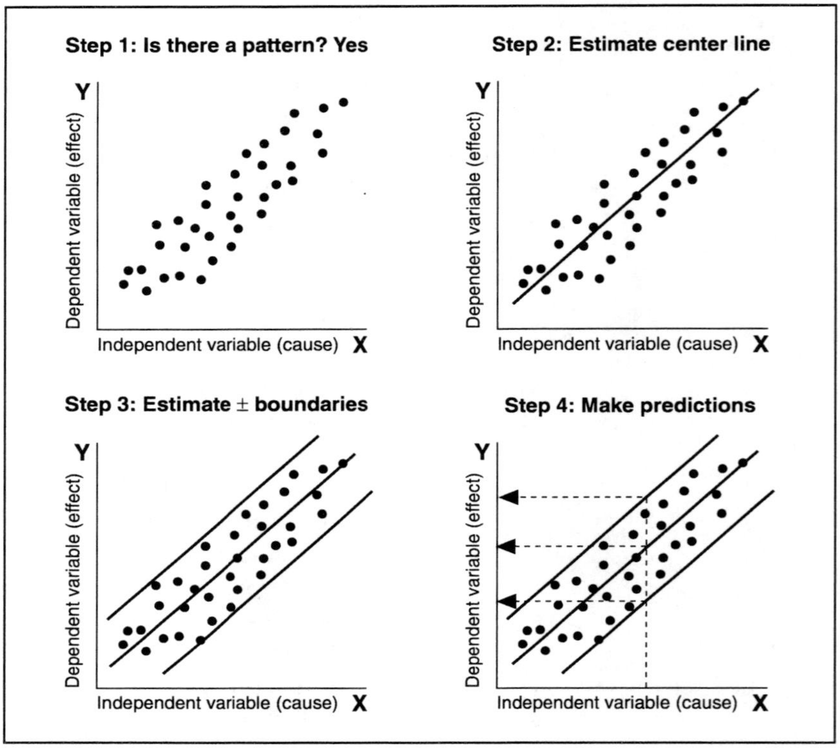

Regression analysis. Regression analysis is the mathematical way of drawing the line and is based on the following formulas:

Formula for the regression line: $y = ax + b$

where "a" is the slope of the line, and "b" is the y-intercept of the line.

Formula for determining "a": $a = \frac{\Sigma xy - n\bar{x}\bar{y}}{\Sigma x^2 - n\bar{x}^2}$

Formula for determining "b": $b = \bar{y} - a\bar{x}$

where "Σxy" is the sum of all the xy combinations; "Σx^2" is the sum of all the x's squared; "n" is the sample size; and "$\bar{x}$" and "$\bar{y}$" are the averages of x and y, respectively.

In order to determine the strength of the relationship, it is necessary to find some measure of correlation. The most usual method of doing so is with the "correlation coefficient."

The correlation coefficient is usually indicated with the letter "r." The correlation coefficient is measured on a scale from –1 through zero to +1. The closer the correlation coefficient is to one (either plus or minus), the stronger is the relationship. The closer the correlation coefficient is to zero, the weaker is the relationship.

As a general rule of thumb for interpreting the strength of the correlation coefficient, it can be said that:

–0.75 to –1.00, 0.75 to 1.00 = extremely strong correlation
–0.50 to –0.75, 0.50 to 0.75 = strong correlation
–0.25 to –0.50, 0.25 to 0.50 = fair to negligible correlation
0.00 to –0.25, 0.00 to 0.25 = weak, if any, correlation

The correlation coefficient (r) can be calculated using the following formula:

$$r = \frac{\Sigma xy - n\bar{x}\bar{y}}{\sqrt{(\Sigma x^2 - n\bar{x}^2)(\Sigma y^2 - n\bar{y}^2)}}$$

The strength of the relationship can also be estimated by looking at the scattergram and seeing how close the plotted points are to the eyeballed line. In many cases, simply eyeballing it is good enough to give you an idea of the nature and extent of the relationship.

While some of the formulas might seem difficult, they are built into most SPC software packages, as well as being available in many popular spreadsheet programs. Also, many hand-held statistics calculators have these formulas built right in. Looking back at our training example, let's look at the scatter diagram with the regression line plotted, formula of the line given, and correlation coefficient given. (See the illustration at the top of page 126).

This can be interpreted to mean that, given this sample data, a firm can expect to have approximately 12.509% defective (this is the y-intercept in the formula) with zero hours of training. Using the formula $y=ax+b$ to find a corresponding % defective, you would substitute the number of hours of training for x in the formula. For each additional hour of training, the firm can expect to see its % defective product reduced by 0.179. Thus, a firm with 10 hours of training would expect to have 10.719% defective product [12.509 – (0.179 × 10)]. A firm with 50 hours of training would expect

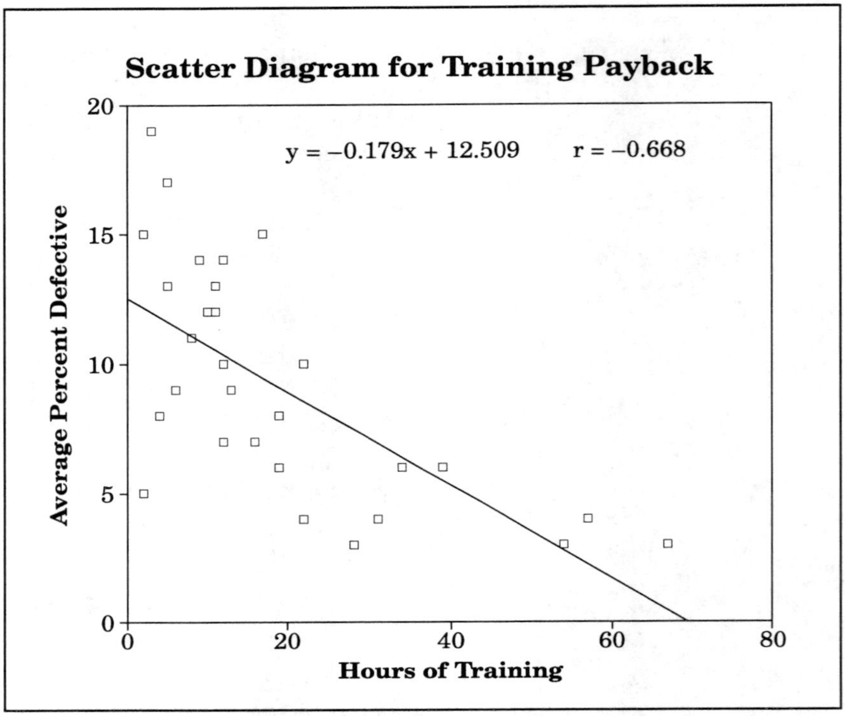

to have 3.559% defective product [12.509 – (0.179 × 50)]. How certain can the firm be of this payback? Our sample indicates a correlation coefficient of –0.668. This is a fairly strong relationship. Thus, the firm can be sure to have some benefit from its training, although it might be somewhat less or more than the equation given in the regression line.

Regression analysis at Casey Press. You will recall that the largest area of opportunity identified in the Pareto analysis was "equipment in poor condition." In a subsequent brainstorming session of the group as to possible solutions to the equipment condition situation, one of the key points to come out was the overall lack of preventive maintenance at Casey Press. Casey was willing to admit that this was true; she had always resisted investing much time and energy in preventive maintenance because it seemed to be too much of an interruption of the normal production process. However, Casey is an open-minded manager and is willing to change, as long as someone can provide the proof to show an adequate return on her investment. So Casey decided to conduct a little survey of her own.

She asked 30 printing organizations the following two questions: (1) "What percentage of your nonproductive time is spent doing preventive maintenance?"; and (2) "What percentage of your nonproductive time is spent in a breakdown mode?" The data that Casey obtained from this survey is given in the following table.

Shop	Percent of Nonproductive Time in Preventive Maintenance	Percent of Nonproductive Time in Breakdown Mode
1	1	20
2	7	5
3	2	15
4	5	12
5	4	12
6	7	11
7	9	9
8	10	10
9	3	16
10	5	14
11	2	18
12	3	17
13	7	14
14	9	10
15	3	16
16	2	15
17	6	13
18	10	8
19	4	16
20	7	13
21	9	12
22	10	4
23	3	17
24	5	16
25	2	20
26	3	18
27	7	13
28	9	12
29	3	17
30	12	3

The next step was to put this data into the form of a scatter diagram and conduct a regression analysis of the data.

First of all it would be necessary to decide which is the independent variable and which is the dependent variable. In this case, it is assumed that increasing levels of preventive maintenance cause decreasing levels of breakdown downtime. Thus:

independent variable = preventive maintenance
dependent variable = breakdown downtime

Knowing this, it is then possible to plot our paired samples and conduct our regression analysis.

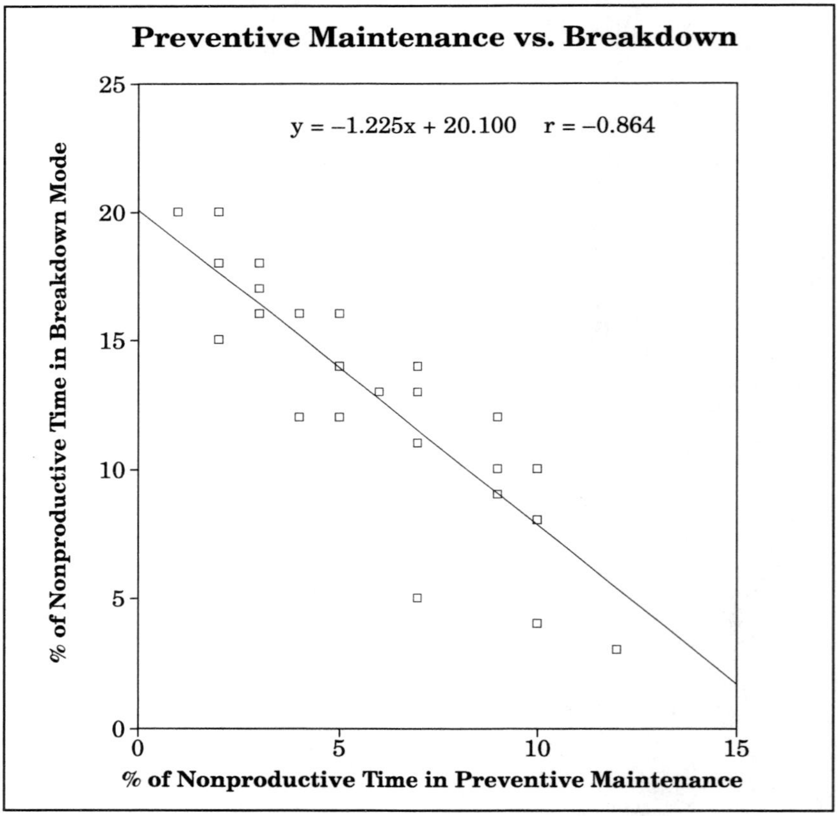

Thus, with no preventive maintenance, a firm can expect a theoretical 20.1% of its nonproductive time to be in a breakdown mode. For each percentage point increase in preventive maintenance as a portion of nonproductive time, the firm can expect to realize a 1.225 percentage point reduction in breakdown mode as a portion of nonproductive time. In other words, for each hour invested in preventive maintenance, the firm will realize a 1.225 hour reduction in breakdowns for an overall reduction in downtime of 0.225 hours. With a correlation coefficient of –0.864, this can be interpreted as virtually a sure thing. Casey was convinced by this study and decided to install a preventive maintenance program as one of her first quality improvement steps. She also instituted a monitoring system to evaluate return on investment.

Histograms and Control Charts

There are two more tools to look at in the quality improvement tool box. These are the histogram and the control chart. Each of these tools has a separate chapter (histograms in Chapter 6 and control charts in Chapter 7); therefore, they will only be introduced here.

In explaining the theory and application of histograms and control charts, it is very important to keep in mind that a control chart is used for determining whether a process is stable and predictable over time. If it is, then it is said to be "in control." A process that is not in control is said to have "special" (also called assignable or sporadic) causes of variation. One must first address these "special" causes and bring the process under control *before* addressing the issue of process capability, which is best analyzed using the histogram. Process capability addresses the relationship between process variation and specifications. If the process varies beyond the specifications, it will invariably produce product that is "out of spec." The process is said to be "incapable."

Therefore, the following steps apply:

1. Is the process in control? Evaluate this using a control chart.
2. Bring the process under control—remove assignable causes.
3. Is the process capable? Evaluate this using a histogram and the statistics of the normal distribution.

It is critical to remember that process control must be evaluated before looking at process capability. Otherwise, the process may be capable one minute, incapable the next, due to the fact that it is not in a stable state. However, and this is a big however, in this book we will be explaining process capability before explaining process control. Why? The answer is that process control and process capability are both based on the theory and application of the normal distribution of data, but it is easier to understand normal distribution using histograms to calculate statistics such as the average and standard deviation. Keep in mind throughout that this is being done only to assist in the explanation. In the real world, we must evaluate process control and bring the process into a statistically controlled state before evaluating process capability.

Histograms and control charts are related to each other statistically, and an understanding of one makes it easier to understand the other. The histogram gives us a picture of the amount of variation in a process during a predetermined period of time. We can use the histogram to tell us what the average is for our process and also to determine what the plus-or-minus is for our process. If we compare the average and plus-or-minus of our process to the expectations we have for that process (specifications), then we are evaluating the capability of our process. This is known as the process capability study.

A histogram can be constructed for any variable you wish to measure. For example, you might wish to look at on-time delivery. In doing so, you will want to select a sample period of time that gives you at least 30 pieces of data. In this case, it might be one month. Next you will measure each job in terms of days plus-or-minus from the delivery date. Then you can put this data into a chart called a frequency diagram, which shows for each range of measurements how many pieces of data there are. Finally, you can determine if this pattern is acceptable in terms of your expectations for the process, which might be plus-or-minus two days from the promised date of delivery. If your variation exceeds the specifications, then you know that the process isn't capable at this point in time of meeting your expectations. Something will have to change.

Control charts are tools for monitoring the performance of a process over time to determine whether it is in control or out of control. "In control" means that the process is stable and therefore relatively predictable. There isn't anything out of the ordinary going on. "Out of control" means that something is amiss in the process. For example, you could also apply a control chart to look at the situation with on-time deliveries. Each job could be logged into the control chart and subgrouped with other jobs. These would be plotted over time to evaluate whether the process is stable, getting worse, or getting better.

Used together, the process capability study (see Chapter 6) and the control chart make an ideal way of monitoring how you are doing over time and if, when you make a change, there is in fact any improvement. If there is, you know that you are on the right track. This is the ***check*** portion of PDCA, in which you evaluate the effectiveness of your changes and take action based on the results.

In the case of on-time delivery, let's assume that we looked at a one-month period of time and found the following:

Job	Days late (–) or early (+)
1	–5
2	2
3	0
4	2
5	1
6	3
7	–1
8	0
9	–4
10	2
11	–3
12	1
13	–3
14	1
15	–2
16	–3
17	–1
18	0
19	–2
20	1
21	0
22	5
23	–1
24	0
25	2
26	4
27	–1
28	–2
29	0
30	2

The next thing to do is to create a "histogram" of this data. Remember again that we normally evaluate process control *first;* it is only for purposes of understanding the statistics that we reverse the order here. A histogram is a bar chart of the items being measured, put in order from lowest to highest, in terms of the number of times each one occurs (frequency) for a given range of measurements. In the diagram on page 133, we show the "on-time delivery" data put into the histogram format. In analyzing the histogram, it is recommended to follow these steps:

1. Evaluate the shape of the histogram to determine if it is it bell-shaped. If it is bell-shaped, we can calculate statistics like mean (average) and standard deviation to easily tell us about the data.
2. What is the centering of the distribution? Where are most of the data situated? This tells us what to expect to occur most frequently.
3. What is the width of the distribution (dispersion or spread)? This tells you the plus-or-minus of your system under normal circumstances.
4. Calculate normal distribution statistics (mean and standard deviation).
5. Compare against specifications.

The following formulas are used for calculating the mean and standard deviation. Most statistics calculators can do this function easily.

Mean (average): $\bar{x} = \frac{\Sigma x}{n}$

Standard deviation: $s_1 = \sqrt{\frac{\Sigma(x - \bar{x})^2}{n - 1}}$

where "Σ" means "sum of everything that follows," "n" means "sample size," and "x" means "individual piece of data."

Using the process capability study at Casey Press. The example of on-time delivery is what the employees at Casey Press actually pursued. They accumulated the data given above over a two-week period of time. They did this because even though there had only been a small number of jobs where "lateness" had been identified as a problem area in the Pareto analysis, they also knew that lateness would in most cases so dissatisfy the customers as to cause them never to return. Thus, this was a critical variable to measure. Additionally, they felt that on-time delivery is one good method for monitoring the overall health of their system (other good

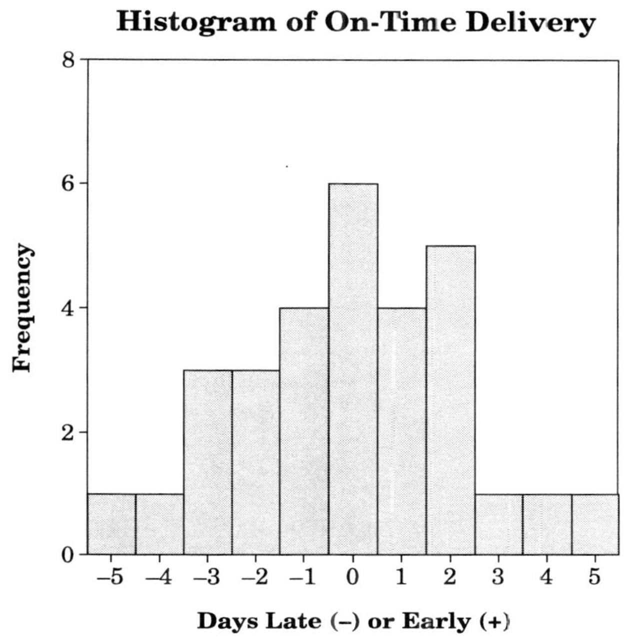

Summary Statistics

Mean	–0.07
Standard Deviation	2.33
Minimum	–5.00
Maximum	5.00
Range	10.00
Kurtosis	–0.51
Skewness	–0.02

ways could include cost of quality, percent defective product, and total cycle time).

Analyzing the data showed that the distribution was, in fact, bell-shaped or "normal." Many of the deliveries fall in the range of 0 days ±1 day, which is the average of the distribution. The width of the distribution is ±5 days total from on-time. Some of the summary statistics for the data are as follows: the average or mean (–0.07) is in fact very close to zero or on-time; the standard deviation, which indicates the plus-or-minus range for 68.5% of the data, is 2.33. Kurtosis (which refers to abnormal concentration of data in the shoulders or abnormally extended tail areas of the distribution) and skewness (which refers to an abnormal concentration of data to one or the other side of the middle of the distribution) are fully described in the Chapter 6, "The Process Capability Study." For now, we will interpret these numbers as good, and the data as normal. If you multiply the standard deviation times three, you have the plus-or-minus for 99.7% of the time.

Thus, you can expect, given this system, that 99.7% of the time your deliveries will be somewhere in the range of:

0 (rounded off from –0.07) ± (3 × 2.33) =
0 ± 7 (rounded up from 6.99) =
–7 to +7 days

...and 68.5% of the time your deliveries would be in the range of the mean plus-or-minus the first standard deviation, or:

0 (rounded off from –0.07) ± 2.5 (rounded up from 2.33) =
–2.5 to +2.5 days

What is the specification? This is what's acceptable to your customers. In commercial printing, ±7 or even ±2.5 days would probably be unacceptable performance. Even though most of your deliveries are close to on-time, there are too many that aren't. At Casey Press, the specification was determined to be on-time ±1 day. Thus, you would conclude that something in the system would have to be changed to reduce the variation in terms of delivery performance. This could include items such as a different scheduling system or better job planning. Looking back at their flow chart, the employees at Casey Press decided that a good way to address the problem was to institute a checklist system, whereby each department would thoroughly "QC" their product before sending it on to the next department. This would reduce the likelihood of product being late due to a final QC rejection. It would also improve the ability of scheduling to provide accurate schedules, since it wouldn't have to factor in the possibility of a job being rejected. Also, they decided to institute a thorough job review process, where each new job would be reviewed with the production department before producing it in order to flag and prevent potential problems before going to press. Before doing any of this, however, they knew that they would have to look at the data in the form of a control chart as well.

Control charts. The control chart is, as previously stated, a tool for monitoring a process over time. We generally use the control chart before using the process capability study since the process needs to be stable to be reliably capable over time. It tells the user whether the process is stable, out of control, or going out of control. The control chart tells you whether to attack special causes of variation (vital few) if the process is out of control or to attack common causes of variation (the system itself) if the process is in control. It can also be used to determine if process improvement efforts have had any success. The mechanics and details of using a control chart are discussed at length in Chapter 7. Our purpose here is to simply introduce the topic.

There are two types of general categories of control charts: (1) variables and (2) attributes. Variables are data based on some sort of a measurement dimension, such as time, distance, weight, density, speed, pH, or temperature. At Casey Press, we are looking at the dimension "days ± from on-time" which is a variable measured in terms of days. Attributes are data based on items that would normally be counted, not measured. Examples of important attributes include % defective product, number of scratches per plate, number of plates made over, number of hickeys, or number of signatures missing staples. It would be very difficult to measure any of these items, so we count them instead.

There are also several types of control charts in each category. The most common types of control charts for variables are the "average and ranges chart" (x-bar/R), the "moving average and ranges chart" (moving x-bar/R), and the "individuals and ranges chart" (x/R). Of these, the x-bar/R chart is by far the most commonly used type. With an x-bar/R chart, the user takes readings of whatever dimension is being monitored. These readings are grouped into what are known as "subgroups." The calculations for the chart are based on the averages and ranges of each of these subgroups, which are then graphed for interpretation.

The most common types of control charts for attributes are the "fraction defective chart" (p), "number of defectives" (np), "count of defects" (c), and "defects per unit" (u). In this terminology, a defect is the individual problem (a hickey, for example), and a defective is a product containing one or more defects (a press sheet with several hickeys, for example). It is merely a question of determining if it is better in your scenario to count defects or defectives.

The basic steps for developing a specific control chart, regardless of type, are as follows:

1. Decide which critical variable you want to monitor (for example, density, dot gain, or register) in analyzing printed sheets; or which attribute you want to count (for example, hickeys, % defective, or plate remakes). It is up to you to decide which variables or attributes to monitor, but you should also follow a few basic guidelines in making this decision:
   - Monitor an item that plays a very important role. Pareto analysis can be used to determine the relative importance of the variable or attribute.
   - Identify the relationship between the characteristic to be monitored and the end product. A scattergram can help determine the existence (or lack of existence) of a correlation between the characteristic and the end product if the characteristic is a variable; a defect cause check sheet might be helpful if it is an attribute.
   - Be sure that the variable or attribute is quantifiable, otherwise it will be of limited diagnostic use and will form little basis for helping to determine what action(s) should be taken.
   - Try to find variables and attributes at the earliest possible point in the process, so they can serve as early warning signals and ensure product uniformity. For example, it is more beneficial to monitor plate remakes than jobs reprinted due to plate problems, more beneficial yet is to monitor plate defects than to monitor plate remakes.
2. Collect initial sample data. Ideally, you will want to collect enough data for at least 20 subgroups before putting together a control chart. If you are forced to use fewer subgroups for starters, then be wary of possible misinterpretation of the chart, since the limits will be only tentative in nature. It is possible to start out with little data; however, you must be wary of drawing conclusions until a larger sample is gathered.
3. Based upon data that you have collected, determine the central line to be used. Generally, this will be the grand average of your data (the average of all subgroup averages). Set control limits. Control limits are usually established based on the statistical calculation of ±3 standard deviations of the subgroups from the center line, thus ensuring that 99.7% of the product will be within the

limits. Warning limits are usually established at ±2 standard deviations from the average. Therefore, if a value falls outside the control limits, then you can say that there are only 3 out of 1,000 chances that this is not due to a change in the process. Points beyond the warning limits will occur naturally only about 5% of the time. In other words, with close to 100% certainty, something in the process has changed, indicating the presence of assignable causes. This all but eliminates the risk of looking for trouble when it doesn't exist.

4. Set up a data collection system. This means that, if the control chart is to be used as a machine-side tool, it should meet the following requirements for simplicity and convenience of use:
   - Simple and error-free (human factor) measurement techniques
   - Instruments must give prompt, accurate, and reliable readings
   - Data should be easy to plot and give the worker immediate feedback; the longer the delay in feedback, the less value it will have in making process adjustments
5. Provide specific instructions on how the results are to be interpreted and what the appropriate actions are to be based upon the results.

In interpreting a control chart, regardless of the type, you are looking for certain patterns to determine whether the process is in control. The most common in-control signs would include the following:

- All points on the chart are within the control limits.
- The points tend to balance each other out on either side of the line of central tendency.
- The points alternate from one side to the other side of the center line, and never continue for more than five points on one side of the line.
- There is a general randomness to the distribution of the points.
- There are no distinct trends heading towards one of the control limits.
- Few points are near the control limits, and no points are outside of the control limits.
- The points do not "hug" the line of central tendency and do not tend to run in straight lines.

Basically, you are looking for a balanced, random pattern. Anything else could spell trouble. A typical, in-control control chart that meets all of the above conditions might appear as shown below.

An In-Control Control Chart

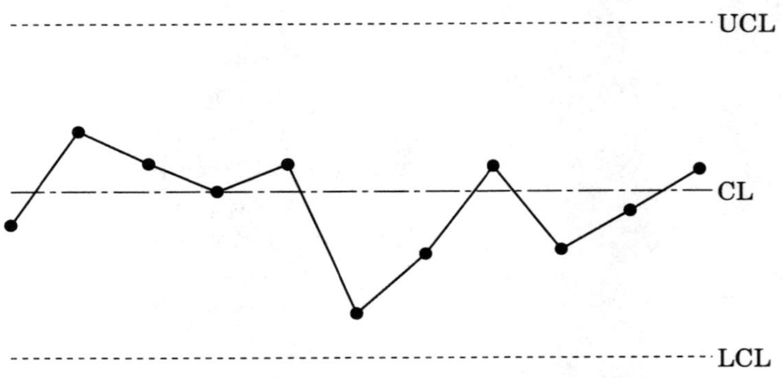

Control charting at Casey Press. As we said, the employees at Casey Press wanted to put the on-time delivery into a control chart before implementing any changes. If the process wasn't in control, they knew that they couldn't trust the results of the process capability study, because their distribution would be subject to unpredictable change. The data from the study is based on a measurement, days late or early relative to the promised delivery date. Thus, they knew that they would have to use a variables control chart. Since the jobs were already in a time sequence, all they had to do was to subgroup them. After some study, they chose to group them in subgroups of three jobs per subgroup. This decision was based on the fact that Casey Press averaged three deliveries per day almost 90% of the time. Thus, there was a logical basis for choosing a subgroup size of 3. However, if Casey Press's delivery schedule varied a great deal, it would be more appropriate to choose a chart such as an individuals' chart where subgrouping isn't a consideration. The key point here is that subgroups must be based on logical rationale and be relatively similar or homogenous, otherwise they aren't rational subgroups but rather arbitrary constructs. The resulting control chart is shown on page 139.

There are actually two graphs involved with an x-bar/R chart. The top chart plots the averages of each subgroup,

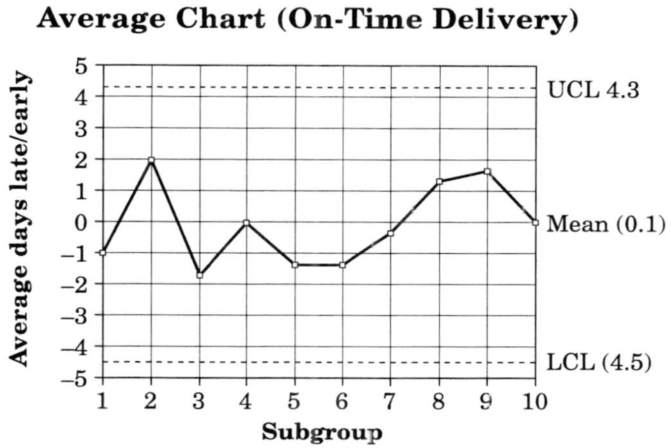

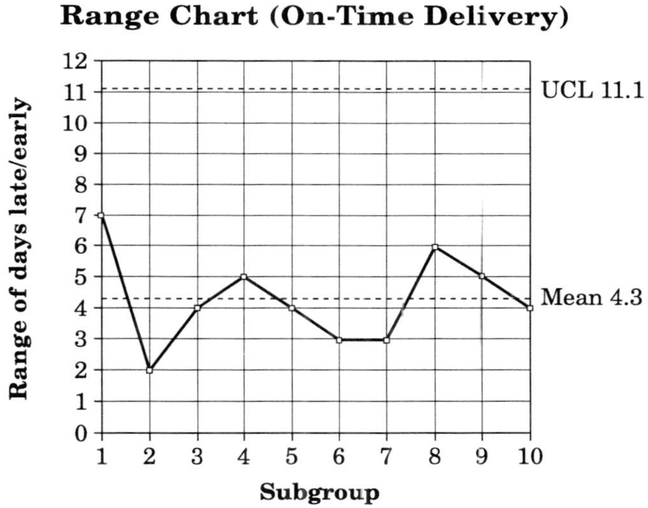

Data

Subgroup	Reading A	Reading B	Reading C
1	–5	2	0
2	2	1	3
3	–1	0	–4
4	2	–3	1
5	–3	1	–2
6	–3	–1	0
7	–2	1	0
8	5	–1	0
9	2	4	–1
10	–2	0	2

thus indicating the average performance of the process. The bottom chart plots the range of each subgroup, thus giving an indication of the level of variation in the process at a given moment in time. Overall, the process can be said to be in control, which is a good indication that the process capability study results can be trusted. Casey Press's team decided to put in place a control chart for on-time delivery that would be posted and updated daily so that everyone would know where they stood. As they improved, they would periodically revise the control chart's limits as necessary. Control limits should be recalculated and revised when, and only when, a changed condition has occurred and the cause of the change has been identified. Otherwise, the condition is not changed but rather can be said to be out of control, which is not a basis for revising limits.

Summary

This chapter has reviewed the most commonly used tools of statistical process control and presented them within the framework of the Plan-Do-Check-Act cycle for scientifically approaching continuous improvement. Individual improvement projects might be narrower or wider in scope, but would certainly draw upon these tools.

6 The Process Capability Study

Capable Processes and the Process Capability Study

In this chapter, we examine a tool called the ***process capability study*** (PCS). It is one of the most common and useful tools in the statistical process control tool kit (in the form of the "histogram"). The PCS consists of the following:

- Collecting data on a process
- Deciding if the process is in a state of statistical control
- Comparing the natural output of the process to the man-made specifications imposed upon it
- Making predictions based upon these findings

Note from the above that process capability assumes a state of statistical control. The concept of capability is discussed in this book before process control only because the statistical principles upon which both are built are more easily explained in conjunction with the process capability study. However, before applying the procedures in this chapter, one must first understand and apply the procedures in the chapter on control charts (chapter 7).

A capable process is a process that will produce virtually all of its product within designated specifications. The information yielded from the PCS is invaluable when analyzing the acceptability of process variation, and determining the status of the process relative to the specifications.

First, to assure that the collected data sufficiently reflects the normal variability in the process, the time frame for the collection of the data must first be examined. Two ways to collect data are:

- In a continuous stream
- Over a period of time

Data collected in a continuous stream will yield information about the short-term variability of the process. Continuous stream data collection is the preferred method for the PCS, because data collected over longer periods of time tends to introduce different subprocesses into the main process.

For example, a printing company might want to investigate the capability of a variable in the bindery department (e.g., trim variation). Examining 100 continuous books will result in a good picture of the short-term variation of the process. However, collecting 100 books over a period of three days would add the variability of multiple shifts to the variable of trim variation, which is what the study is centered on. If one shift operates differently than the next, we are essentially looking at two different processes, each with its own inherent variability.

Additionally, looking at other factors is necessary to ensure that data only comes from a single source. Other factors that might be considered include the following:

- Is the output from two different machines being measured?
- Do different data collectors (people or instruments) give different results?
- Are incoming materials (ink, paper, etc.) coming from different sources?

All of these questions must be asked, and answered, in order to be certain that the PCS is actually measuring the capability of a single process.

Second, the data must be statistically sound. Besides assuring that the data only reflect a single process, enough data must be collected to provide a degree of statistical reliability. For example, measuring ink density for two press sheets is not enough to predict the variability of the process, even over short periods of time. Statisticians generally recommend at least 30 and, preferably, 100 individual samples for a PCS. Less data will provide questionable results.

Third, only relevant characteristics must be observed. The characteristics being measured must have some relationship to the quality (either perceived or real) of the product. Methods of determining the importance and relevance of characteristics include:

- ***Cause-and-effect analysis.*** This method is used to come up with a list of important variables.
- ***Pareto analysis.*** This method is used to prioritize quality characteristics according to the magnitude of their contributions to quality variation.

- ***Correlation or scatter diagrams.*** These diagrams will identify the strength and nature of the relationship between a cause and its corresponding effect.

Although the process of identifying relevant characteristics seems almost common sense, a large number of quality initiatives have failed in its absence. A major error made by many, especially in the beginning stages of the PCS, is the assumption that a quality characteristic is important, when it is, in fact, irrelevant. Taking the time initially to examine and question all underlying assumptions is one of the most important first steps in any PCS.

An Example to Illustrate the PCS

Our fictional printing company, Casey Press, was concerned about its perfect-bound books falling apart. Through cause-and-effect analysis, employees at Casey Press determined that many variables related to the glue in the spines of these books could have an effect on the overall strength of the binding. Among these were glue pot temperature, plasticizers, glue film thickness, mixing time, and book thickness.

The team members conducted a test on 100 books in order to create a Pareto analysis. Each failure of a book to remain intact during the test was traced back to one of the five variables listed in the previous paragraph. The results of the test showed that 75% of the books that failed had inadequate glue film thickness (GFT).

Finally, they conducted a correlation study using a scatter diagram to determine the statistical strength of the relationship between GFT and binding strength. This study showed that, both at low GFT and at excessively high GFT, the books were more likely to fall apart. However, between a certain range of thicknesses almost no books fell apart. The acceptable range was found to be 0.005±0.002 in. The study also revealed that the center of this range was optimal; therefore, "0.005±0.002 in." became the specifications or tolerances for the process. In addition, the correlation study showed that 90% of the failures could be attributed to the relationship between either high or low GFT and books falling apart.

After deciding which variable to monitor (GFT), the company began collecting data to evaluate the capability of the process. A total of 32 samples were taken continuously, and the GFT was measured in thousandths of an inch, giving the following data:

Sample	GFT*	Sample	GFT*	Sample	GFT*
1	5	11	5	21	4
2	3	12	8	22	5
3	5	13	8	23	5
4	3	14	6	24	6
5	5	15	5	25	5
6	4	16	5	26	5
7	2	17	5	27	5
8	2	18	6	28	3
9	7	19	4	29	5
10	6	20	6	30	4
				31	4
				32	4

*Glue film thickness in thousandths of an inch.

Deciding if the Process Is in Statistical Control

After collecting the data, the next step in the PCS is to decide if the process is in a state of statistical control. Various definitions exist both for statistical control and for the methods for evaluating statistical control. The following definition and method of evaluation best fulfills the requirements of the vast majority of cases in any printing company.

Statistical control means that the process is predictable and varying in a purely random fashion over time. The easiest way to evaluate whether a process is in control is to draw a frequency diagram (histogram) of the data from the process, and see if the histogram forms the bell-shaped curve characteristic of the normal distribution. There are, however, cases where a process can produce a normal distribution and at the same time be out of control. You will, therefore, want to put the data into the format of a control chart to analyze process control (see Chapter 7 on process control). To construct the histogram, divide the full range of the data into several cells and draw a bar chart that shows the frequency of data in each of these cells. In the example of Casey Press, the lowest value is 0.002 in. and the highest value is 0.008 in. In the following chart, this range of 0.006 in. is divided into cells with a width of 0.001 in., and the frequency of observations in each cell is tallied:

Cell	Tally	Frequency
0.002	xx	2
0.003	xxx	3
0.004	xxxxx x	6
0.005	xxxxx xxxxx xxx	13
0.006	xxxxx	5
0.007	x	1
0.008	xx	2
	TOTAL	**32**

Displayed as a histogram, this same information would appear as follows:

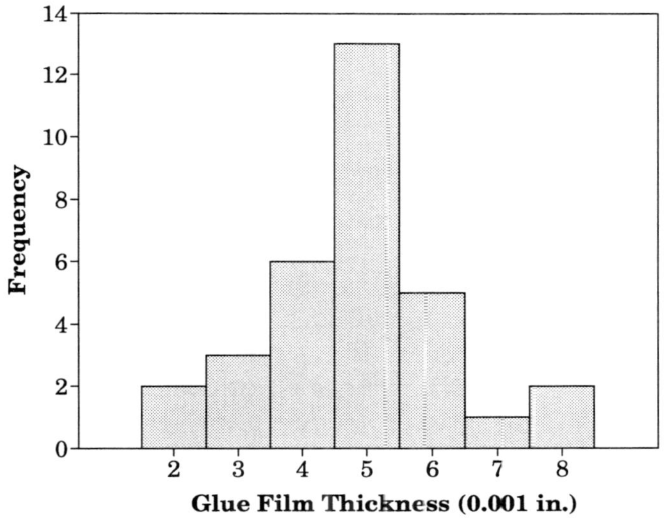

The shape shown in the above histogram is typical of a process following a normal pattern of variation. The data is centered near 0.005 in. and decreases towards the ends of the distribution.

The two general requirements for a normal distribution of data are:

- Symmetry about the center point (mean)
- Concentration about the mean with decreasing data in the tails

Other distributions, which are not considered to be normal, are:

- Distributions displaying skewness
- Distributions displaying kurtosis
- Distributions that are multimodal or random

Distributions displaying skewness. Skewness consists of an abnormal concentration of data to one or the other side of the middle of the distribution. In the normal curve, the mean and the median are both at the center of the distribution with a skewness value of 0. However, in a skewed curve, the mean is either to the left (positive skew), or the right (negative skew) of the middle point.

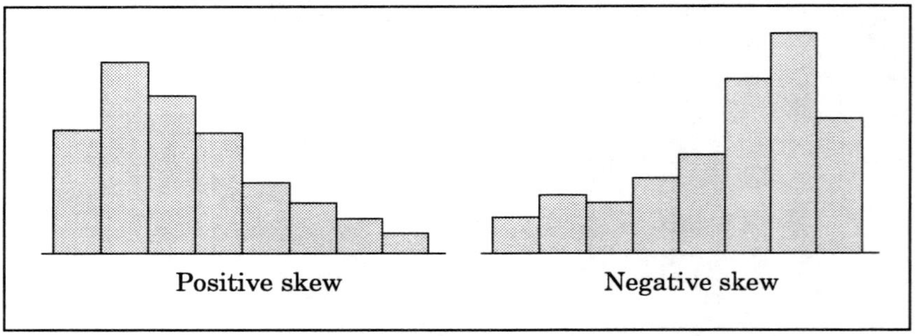

Distributions displaying kurtosis. Kurtosis refers to abnormal concentration of data in the shoulders or abnormally extended tail areas of the distribution. A normal curve will exhibit 68.5% of its data within ±1 standard deviation of the mean. If more or less of this amount exists, the curve will either have a flat peak, fat shoulders, and less tails (kurtosis <0) or a sharp peak, thin shoulders, and longer tail areas (kurtosis >0). The normal curve has a kurtosis exactly equal to 0.

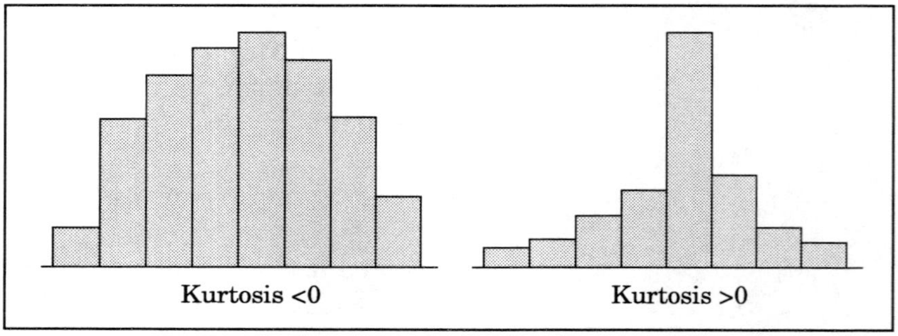

Distributions that are multimodal or random. The diagram on the left below indicates two sets of data in the distribution (bimodal). To evaluate the data, it would be necessary to separate (stratify) the data by process. For example, maybe the readings are being taken using two different instruments or the job is being produced on two different presses. The right distribution appears to be completely random, thus indicating complete unpredictability. In practice, distinguishing between these two types of distributions may be difficult. Multiple processes are often difficult to distinguish. Therefore, it is very important to plan carefully to eliminate the possibility of incorrect sampling from multiple processes.

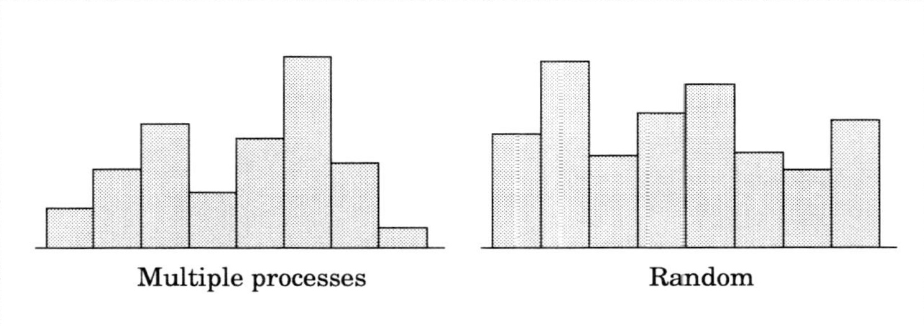

Although mathematical formulas exist for handling skewness and kurtosis, they are difficult and complex. However, a good SPC software package or even a good spreadsheet program will contain provisions for handling these formulas and other types of non-normal data. Additionally, a good SPC software package should perform statistical tests for normality at various levels of confidence (lets you know to what degree you can depend upon your data). Let us now apply these concepts to the PCS data for Casey Press.

Referring once again to the histogram from Casey Press (page 145), the computer tells us the following information: (1) the data is normal with 95% confidence, (2) the skewness is 0.14, and (3) the kurtosis is 0.08.

The conclusions from this would be that:

- The mean of the data is slightly to the left of the median point.
- The distribution is slightly more peaked than the pure normal distribution.
- Neither of these conditions is extreme enough to indicate a non-normal distribution.

Comparing the Natural Output of the Process to the Man-Made Specifications Imposed upon It

The next step is to determine what the natural process output is, and then compare this output with the predetermined specifications. To determine the natural output of the process, we must calculate the mean and the standard deviation for the process. The mean will be the arithmetic average and will tell us where the process is centered. The standard deviation will tell us how much variability can be expected from the process. In general, the number of items located between the mean and the standard deviation is a constant, where:

$$\text{mean} \pm 1\text{s} = 68.5\%$$
$$\text{mean} \pm 2\text{s} = 95\%$$
$$\text{mean} \pm 3\text{s} = 99.7\%$$

From this information, the conclusion can be drawn that virtually all measurements (99.7%) will normally fall somewhere between the mean and ±3 standard deviations from the mean. Therefore, the region of the mean ±3 standard deviations is considered the natural process output.

The following formulas are used to calculate the mean and standard deviation. In these equations, "x" is a single piece of data, "Σ" (capital "sigma") is the sign for summing up all of the data, and "n" is the size of the sample under investigation:

$$\textbf{\textit{Mean (average):}}\ \bar{x} = \frac{\Sigma x}{n}$$

$$\textbf{\textit{Standard deviation:}}\ s_1 = \sqrt{\frac{\Sigma(x - \bar{x})^2}{n - 1}}$$

We define the first standard deviation as "s sub-1" (s_1) in order to differentiate it from the second and third standard deviations. From our 32 pieces of data, we get the following results:

Mean = 4.844
Standard deviation = 1.417

Next we must calculate the third standard deviation in order to find out what the 99.7% boundaries of our process are. This is done by multiplying the first standard deviation by three. Thus:

Third standard deviation = (3)(1.417) = 4.251

From these calculations, we can conclude that the natural output of the process will be:

$$4.844 \pm 4.251$$

This means that 99.7% of the time, the process will produce data within the range of 0.593 (4.844 – 4.251) to 9.095 (4.844 + 4.251). How well does this compare to our specification of 5±2? The following histogram, with process limits based on ±3 standard deviations from the mean and specification limits based on 5±2, gives us the answer.

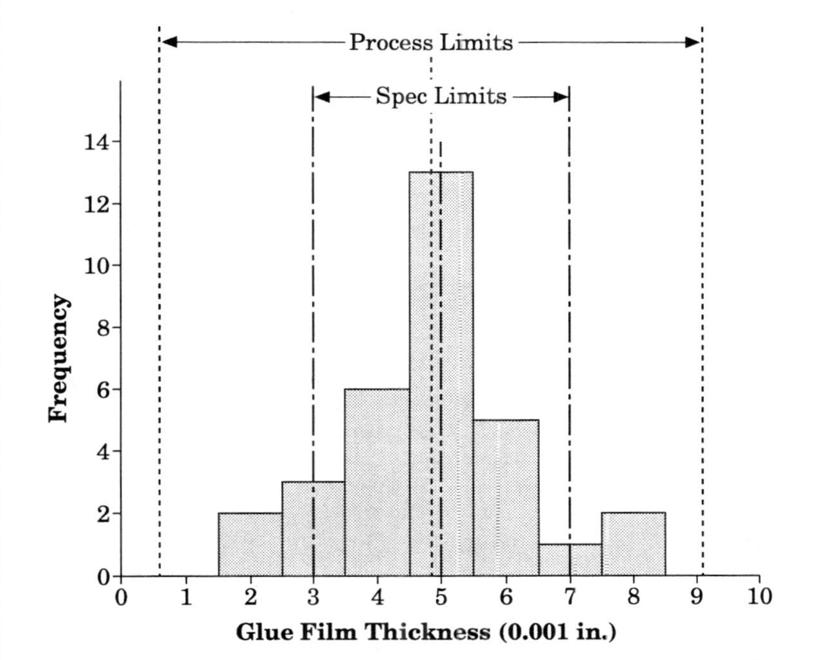

As the histogram shows, the actual limits of the process are beyond those of the specification. Thus, some portion of the process's output will be out of specification. Since the process seems to be well centered on the specification's nominal dimension (0.005 in.), we can conclude that the real problem is the excessive variability of the process.

Process Capability Indices

Using an index to describe to what extent a process is capable, or incapable, is quite common. These indices will be used as benchmarks for monitoring quality improvement over time. We, of course, do not need an index to tell us that our process is incapable. When the ±3 standard deviation limits fall beyond one or both specification limits, the process is deemed incapable, and our example falls beyond those limits.

However, there are varying degrees of process capability. For example, a process might be centered exactly on the nominal dimension of the specification and have the same process limits as the plus-or-minus values of the specification limits. As long as this process stays centered and the variation doesn't increase, there will be no out-of-specification product. But as soon as anything causes the process to shift, there will be out-of-specification products. Thus, having a process centered on its specification and well within the specification limits is desirable. The following illustrations demonstrate the relationship between a process's output and its specification limits in terms of the capability of that process to produce a satisfactory product:

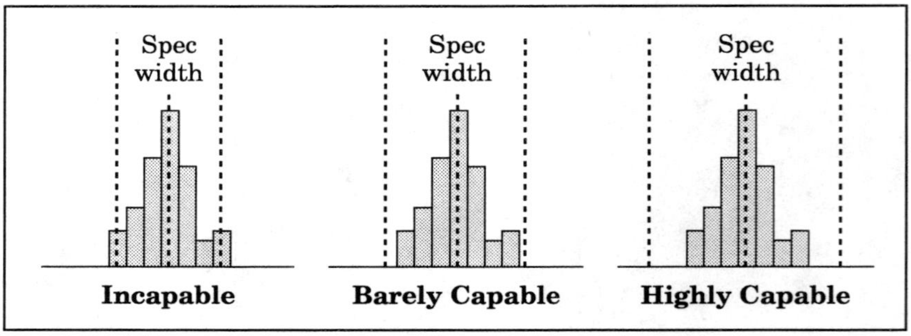

The incapable process will always produce some out-of-spec products. Thus, our process virtually guarantees that books will fall apart on a consistent basis. The barely capable process will produce good products only as long as there is no shift in its average output or increase in its variability. The highly capable process will produce products within specification, even if there is some shift in its center or increase in its variability.

The process capability indices show these relationships mathematically. The two most common indices are "Cp" and "Cpk":

- ***Cp index*** is defined as the ratio of the tolerance width (specifications limits) to the process width. It is calculated as follows:

$$Cp = \frac{[USL - LSL]}{6(s_1)}$$

where USL and LSL are "upper specification limit" and "lower specification limit," respectively, and "s-sub 1" is the first standard deviation.

- If Cp is greater than one, the process is capable. The goal is the highest possible Cp. The problem with Cp is that it says nothing about how well the process is centered on its specification.
- ***Cpk index*** is defined as the ratio of the distance between the mean and the nearest specification to the mean to one-half the process width. It is calculated as:

$$Cpk = \frac{[\bar{x} - NSL]}{3(s_1)}$$

where "NSL" is the "nearest specification limit," and the difference between the average and NSL is taken as an absolute value.

- Cpk indicates both the ratio of process width to spec width and process centering on the specification.

A common goal for the Cpk index is 1.33, which is equivalent to an exactly centered process with a capability of 8 standard deviations. If the process isn't exactly centered, Cpk will always be less than Cp. Taken together, these two measures can give an excellent picture of process capability.

In the case of our example, the indices are calculated as follows:

$$Cp = [7 - 3] / [6(1.417)] = 0.47$$

$$Cpk = [\,|4.844 - 3|\,] / [3(1.417)] = 0.43$$

These calculations indicate, as we have already seen, that the process is not capable since it is both too variable and a little off center.

Remember that it is imperative to determine if the process is in control (utilizing a control chart) before making process capability calculations. The process capability indices will invariably be incorrect if they are calculated for a process that isn't in a state of control.

Making Predictions Based upon the Findings

The final step in the process capability study is to make predictions based on the findings from the study. To accomplish this step, it is necessary to use the statistical principles of probability.

Probability is used to determine the likelihood of a random event occurring. In the PCS, probability can be used to predict what percentage of the product is likely to be out of spec, given the process and its specification limits.

It is possible to predict probability with the use of the normal distribution. The areas underneath the curve represent the probability of occurrences happening at a given point or in a given range. Since all normal curves share the same characteristics in regards to their shape and area, statisticians have designed what is known as the ***standard normal distribution.*** The standard normal distribution has areas underneath its curve to which all normal curves can be related in order to predict probability.

Probability, as it relates to quality control, is interested in occurrences happening within or beyond a range of values. The probability of a specific occurrence is of little interest. In other words, it is of little value to know how many books will be produced with a GFT of exactly 0.003 in. Rather, our interest lies in knowing how many books will be produced within and/or beyond a range of values. The range of values of importance is, of course, the specification limits.

Probability is usually expressed as a value between zero and one:

0.00 = 0% probability of an occurrence
0.50 = 50% probability of an occurrence
1.00 = 100% probability of an occurrence

The probability of an event that is certain is equal to one, and the total area under the normal curve is also equal to one. That is, there is a 100% probability of an occurrence happening somewhere underneath the curve.

In order to facilitate the calculation of probabilities using the normal curve, tables have been set up for the standard normal distribution showing the areas between the mean and some value of interest. Here the value of interest is either the upper or lower specification limits since we are concerned with the percentage of product that would be out of spec on either end of the production range. One such table is shown on the next page.

To use these tables the following information is required:

- Mean of the distribution
- Standard deviation (s) of the distribution
- The value of interest (x)

Areas Underneath the Curve of the Standard Normal Distribution

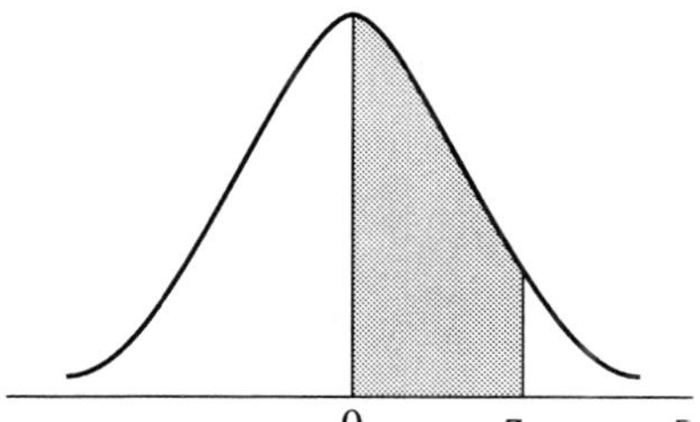

z or $\frac{x-\bar{x}}{s_1}$	.00	.01	.02	.03	.04	.05	.06	.07	.08	.09
0.0	0.00000	0.00399	0.00798	0.01197	0.01595	0.01994	0.02392	0.02790	0.03188	0.03586
0.1	0.03983	0.04380	0.04776	0.05172	0.05567	0.05962	0.06356	0.06749	0.07142	0.07535
0.2	0.07926	0.08317	0.08706	0.09095	0.09483	0.09871	0.10257	0.10642	0.11026	0.11409
0.3	0.11791	0.12172	0.12552	0.12930	0.13307	0.13683	0.14058	0.14431	0.14803	0.15173
0.4	0.15542	0.15910	0.16276	0.16640	0.17003	0.17364	0.17724	0.18082	0.18439	0.18793
0.5	0.19146	0.19497	0.19847	0.20194	0.20540	0.20884	0.21226	0.21566	0.21904	0.22240
0.6	0.22575	0.22907	0.23237	0.23565	0.23891	0.24215	0.24537	0.24857	0.25175	0.25490
0.7	0.25804	0.26115	0.26424	0.26730	0.27035	0.27337	0.27637	0.27935	0.28230	0.28524
0.8	0.28814	0.29103	0.29389	0.29673	0.29955	0.30234	0.30511	0.30785	0.31057	0.31327
0.9	0.31594	0.31859	0.32121	0.32381	0.32639	0.32894	0.33147	0.33398	0.33646	0.33891
1.0	0.34134	0.34375	0.34614	0.34850	0.35083	0.35314	0.35543	0.35769	0.35993	0.36214
1.1	0.36433	0.36650	0.36864	0.37076	0.37286	0.37493	0.37698	0.37900	0.38100	0.38298
1.2	0.38493	0.38686	0.38877	0.39065	0.39251	0.39435	0.39617	0.39796	0.39973	0.40147
1.3	0.40320	0.40490	0.40658	0.40824	0.40988	0.41149	0.41309	0.41466	0.41621	0.41774
1.4	0.41924	0.42073	0.42220	0.42364	0.42507	0.42647	0.42786	0.42922	0.43056	0.43189
1.5	0.43319	0.43448	0.43574	0.43699	0.43822	0.43943	0.44062	0.44179	0.44295	0.44408
1.6	0.44520	0.44630	0.44738	0.44845	0.44950	0.45053	0.45154	0.45254	0.45352	0.45449
1.7	0.45543	0.45637	0.45728	0.45818	0.45907	0.45994	0.46080	0.46164	0.46246	0.46327
1.8	0.46407	0.46485	0.46562	0.46638	0.46712	0.46784	0.46856	0.46926	0.46995	0.47062
1.9	0.47128	0.47193	0.47257	0.47320	0.47381	0.47441	0.47500	0.47558	0.47615	0.47670
2.0	0.47725	0.47778	0.47831	0.47882	0.47932	0.47982	0.48030	0.48077	0.48124	0.48169
2.1	0.48214	0.48257	0.48300	0.48341	0.48382	0.48422	0.48461	0.48500	0.48537	0.48574
2.2	0.48610	0.48645	0.48679	0.48713	0.48745	0.48778	0.48809	0.48840	0.48870	0.48899
2.3	0.48928	0.48956	0.48983	0.49010	0.49036	0.49061	0.49086	0.49111	0.49134	0.49158
2.4	0.49180	0.49202	0.49224	0.49245	0.49266	0.49286	0.49305	0.49324	0.49343	0.49361
2.5	0.49379	0.49396	0.49413	0.49430	0.49446	0.49461	0.49477	0.49492	0.49506	0.49520
2.6	0.49534	0.49547	0.49560	0.49573	0.49585	0.49598	0.49609	0.49621	0.49632	0.49643
2.7	0.49653	0.49664	0.49674	0.49683	0.49693	0.49702	0.49711	0.49720	0.49728	0.49736
2.8	0.49744	0.49752	0.49760	0.49767	0.49774	0.49781	0.49788	0.49795	0.49801	0.49807
2.9	0.49813	0.49819	0.49825	0.49831	0.49836	0.49841	0.49846	0.49851	0.49856	0.49861
3.0	0.49865	0.49869	0.49874	0.49878	0.49882	0.49886	0.49889	0.49893	0.49897	0.49900
3.1	0.49903	0.49906	0.49910	0.49913	0.49916	0.49918	0.49921	0.49924	0.49926	0.49929
3.2	0.49931	0.49934	0.49936	0.49938	0.49940	0.49942	0.49944	0.49946	0.49948	0.49950
3.3	0.44952	0.49953	0.49955	0.49957	0.49958	0.49960	0.49961	0.49962	0.49964	0.49965
3.4	0.49966	0.49968	0.49969	0.49970	0.49971	0.49972	0.49973	0.49974	0.49975	0.49976
3.5	0.49977									
3.6	0.49984									
3.7	0.49989									
3.8	0.49993									
3.9	0.49995									
4.0	0.49997									

Generally, the "value of interest" is the upper or lower specification limit. What you are doing is determining what percentage of the product produced falls between the average and each limit. Knowing this, it is then possible to determine the percent of product that will be out of specification.

Of interest is the percentage of the normal curve between the mean and the value of interest To calculate this percentage, the following formula is used:

$$z = \frac{x - \overline{x}}{s_1}$$

The number found in the table under "z" can be interpreted as the percentage of product on that side of the mean between the mean and the specification limit.

To illustrate the use of this method, the following example draws on the glue film thickness data from Casey Press. First, we will be interested in the percentage of out-of-spec product above the specification. The values needed for the formula are:

- x: the value of interest is the upper spec limit, which was 7
- The mean of the distribution, which was 4.844
- The standard deviation, which was 1.417

Using these values in the formula yields:

$$z = \frac{7 - 4.844}{1.417} = 1.52$$

Next, look up "z = 1.52" in the table of areas underneath the standard normal distribution. This value is found by coming down the first column to 1.5 and then going over to the column headed .02 (thus giving 1.52). The number at that point is 0.43574. As this number relates to only 50% of the curve, to determine how much product will be beyond the z-value, 0.43574 is subtracted from 0.5.

$$0.5 - 0.43574 = 0.06426 \approx 6.4\%$$

In other words, approximately 6.4% of the product that is produced will be out of spec on the high side. Now look at the percent of product that will be out of spec on the low side. The x-value is the lower specification limit, therefore:

$$z = \frac{3 - 4.844}{1.417} = -1.30$$

Looking up the value "z = 1.30" (negative values are ignored in z-calculations) shows a standard area of 0.40320. Again, this is only for 50% of the area under the curve, so the percent out of spec is found by subtracting from 0.5:

$$0.5 - 0.40320 = 0.0968 \approx 9.7\%.$$

Use of these calculations can help to predict how much of a product will be "OK" and how much will be defective. This can be very useful information when trying to determine inspection sampling plans or attempting to come up with a budget for quality improvement.

Thus, using the process capability study, we have shown that a printing company can:

1. Determine whether the process is in a state of statistical control.
2. Demonstrate whether that variation is excessive relative to the specifications, thus indicating that there will be out-of-spec products.
3. Determine if a process is capable.
4. Determine that the total amount of out-of-spec product with this process will be approximately 16.1% (6.4% out of spec on the high side and 9.7% out of spec on the low side).
5. Use the information for justifying a process improvement activity such as better materials, equipment, or methods.

Summary

The process capability study is one of the most useful tools in statistical quality control as a tool that can yield a great deal of information on the process under examination. Perhaps the most useful aspect of the PCS is its ability to be used as a benchmark for quality improvement. With effective quality systems in place, printing companies will be able to see their processes become more and more capable of performing to customer needs and requirements. The sales staff will also be able to use the results of the PCS as a selling point for the company's products.

7 Using Control Charts to Monitor the Process

What Is a Control Chart?

A control chart is the best tool you can use for determining whether a process is in control or out of control. As described in Chapter 5, an in-control process is one that lacks "assignable causes" or "special causes" of variation. This means that the process's output will be consistent over time. That is not to say that the process will be capable of meeting your needs or your customer's requirements, just that the results will be relatively consistent.

Attributes vs. Variables

When establishing a program for the ongoing monitoring of quality characteristics, it is necessary to differentiate between attributes and variables, for there are specific types of control charts to deal with different types of quality characteristics.

Variables are those quality characteristics that can be measured on a continuum. Therefore, they are also sometimes called continuous variables. Such characteristics in printing could include ink density, fountain solution pH, and trim variation. Notice that each of these can be monitored by a measurement system.

Attributes, on the other hand, are characteristics that we monitor by counting. Examples of attributes include hickeys, wrinkles, scumming, typographical errors, and missing staples. As you can see, these types of characteristics are either present or not present in the printed product. Therefore, they lend themselves to counting. There are several ways of counting attributes, which will be discussed shortly.

It is important to note that there can be a fine line between attributes and variables. Most attributes can be measured. We could, for example, classify dot-to-dot register as either an attribute or a variable. It is possible to take a magnifier with a built-in scale and measure (in millimeters, inches, etc.) how colors register to each other. Conversely, we could categorize register as an attribute by saying that anything beyond one-half row of dots is not acceptable. Thus, we have a situation that can be monitored by measuring or by a go/no-go process.

The essential questions in determining whether to monitor a characteristic as an attribute or a variable are as follows:

1. Is it economical to measure? Even though most things can be measured, it is often more effort than it is worth. In the case of register, the customer probably evaluates the printed product as either good or bad. To the customer, it is not a question of how much is it out of register, since noticeable misregister is unacceptable. In this case, it would probably be just as effective to monitor register as an attribute.
2. Does the characteristic lend itself to counting or measuring? Some items are very easy to count, while others are easily measured. Scumming is a go/no-go situation that can easily be counted (press sheets with scumming). Density, on the other hand, is a characteristic that can easily be measured with a densitometer.

Disadvantages of Monitoring Attributes

It should be noted that there are a few drawbacks in the selection of monitoring a characteristic as an attribute rather than as a variable.

First, control charts for attributes will yield less information about the process than control charts for variables. The underlying principle for all control charts is that variation is universal and must be reduced in order for product quality to improve. With a variables control chart, we are directly measuring key process variables and can directly observe the cause-and-effect relationship between process variation and product variation. With an attributes control chart, we are monitoring the consequences of excessive process variation. Thus, we are looking at the completed product as a tool for controlling the printing process. Unfortunately, this removes us from the process and makes the feedback loop less effective.

Second, besides making the control chart less effective as a process improvement tool, monitoring attributes might allow the production of defective product. With variables control charts, we can catch the variation before it leads to poor-quality printed product. However, attributes by definition are characteristics of poor quality, which we would rather not see in the product in the first place.

Third, for statistical reasons, we must take very large samples for attributes charts to be valid. Whereas variables control charts can function with small subgroups such as 3 pieces per subgroup, attributes charts necessitate subgroups with a minimum size of 25, and subgroups with a size of 100 are not uncommon. Therefore, unless the data collection system is extremely efficient, there will be significant time lapse between production and control.

These drawbacks aside, there are very good applications for attribute control charts, especially those where measurements are difficult or expensive or where inspection processes are already in place that are capturing attribute data.

The Statistics of Control Charts

Any time when a situation is classified as good or bad, on or off, present or not-present, or go/no-go, it is characterized by the ***binomial distribution.*** However, the control chart concept is based on the characteristics of the ***normal distribution*** with its mean and predictable standard deviations. Fortunately, with large enough samples any distribution will begin to resemble the normal distribution. This is known as the ***central limit theorem*** in statistics—the distribution of a random sample from any distribution, whether binomial, normal, skewed, or even random, will be normal in nature. Thus, by taking samples of large enough size, we can chart attributes in the same way that we chart variables. However, in using control charts, one need not feel compelled to always use large sample sizes. This is due to the fact that we set control limits generally at ±3 standard deviations. With such wide control limits, it is unlikely that an out-of-control sign would go undetected, even if employing a control chart with relatively small samples on a process that isn't normally distributed.

Defects vs. Defectives

The final decision that must be made in choosing which type of attributes chart to use is how do we want to count the attributes. Defects are individual attributes. A hickey is a defect, for example. On the other hand, a defective is a piece of product that contains one or more defects. Thus, a press sheet with hickeys is a defective. Sometimes it is more applicable to count defects, and sometimes it is more applicable to count defectives. The key to determining a counting method is a classification of defects. If it takes several defects to force a product's rejection (minor defects), then it is the defects that we wish to count. If, however, it only takes one or two defects for a product to be rejected, it is preferable to count the defectives. Sometimes it may be advantageous to count both ways and look at the results.

Subgroup Size

The size of the subgroup for an attribute control chart is based on the quality of the process. If the process produces many defects, only a small subgroup is required to detect the lack of control. If the process produces very few defects, a very large subgroup must be used to detect lack of control. As a general rule, if the subgroups don't average at least two to three defectives or defects (depending on which you're counting), the subgroup size must be increased. At some point, if the control chart is being used to improve the process, the process will produce so few defects that extremely large subgroups will be required. At that point, the control chart becomes obsolete. The better the process, the lower the frequency of defects and the larger the subgroup size necessary to the point where the chart becomes uneconomical.

For a variables control chart, the minimum subgroup size is one. That is, you can take a series of single readings and create a control chart out of them. However, it is preferable to have between three and five readings per subgroup because very small subgroups cause a chart to be difficult to interpret; there is simply too much variation. The larger the subgroup, the better the picture you are going to get of your process.

Different Types of Control Charts

We will now look at several types of control charts. For variables, we will look at the ***x-bar/R chart,*** which is the most common type. Other common types include x-bar/s, which is often used with very large subgroups, and moving x-bar/R and individuals charts, which are used for individual readings. We will then look at the two most common types of control charts for attributes: the ***p chart,*** which is used for monitoring percent defective product, and the ***c chart,*** which is used for monitoring the rate of defects in a process. Other common attributes charts include the np chart for monitoring the rate of defectives and the u chart, which is used for monitoring a count of defects.

Monitoring Variables: The x-bar/R Chart

When monitoring variables such as dot gain, pH, or temperature, the most common type of control chart is the x-bar/R chart. Actually, this is two charts in one. The first chart, x-bar, is used for looking at the process average over time. The second chart, R, is used for monitoring the process average variation over time. If you think of the shape of a bell curve, the x-bar chart would tell you where the curve is centered, and the R chart would tell you how wide the curve is. Taken together, the two charts give you an excellent picture of your process results and are excellent tools for discovering when something has or is about to go wrong.

The actual mechanics of putting together a control chart for variables are quite straightforward, although you might want to consider investing in an SPC software package to actually generate the chart, since the arithmetic can be quite time-consuming. The specific steps to follow in putting together an x-bar/R chart are:

1. Decide on a variable to monitor, such as ink density, over time.
2. Establish a system for measuring the variable, such as a densitometer.
3. Determine how frequently to take measurements. This is a rational decision and should be based on how frequently you would expect the process to drift away as well as the volume of the process and how stable the process is over time. The cost of measurement (particularly in terms of time) is often also a consideration.
4. Determine how many items per subgroup each time a measurement is taken. This will be based on an examination of how many readings it takes to get a good picture of the process. For example, with press sheets, between three and five readings per subgroup seems to work well.
5. Take samples for your initial control chart. This should be comprised of at least twenty subgroups.
6. Calculate the average (x-bar) and range (R) for each subgroup. The average is the total of the subgroup readings divided by the number of readings per subgroup. The range is the highest reading in the subgroup minus the lowest reading in the subgroup.
7. Calculate the grand average (x-double bar) and average range (R-bar). X-double bar and R-bar are shown in their symbolic form in the formulas that follow. The grand average is the sum of all subgroup averages divided by the number of subgroups. The average range is the sum of all subgroup ranges divided by the number of subgroups.
8. Calculate the control limits for the x-bar chart using the following formulas:

$$CL = \bar{\bar{x}}$$

$$UCL = \bar{\bar{x}} + (A2)(\bar{R})$$

$$LCL = \bar{\bar{x}} - (A2)(\bar{R})$$

where "CL," "UCL," and "LCL" are the "center line," "upper control limit," and "lower control limit," respectively.

9. Calculate the control limits for the R chart using the following formulas:

$$CL = \bar{R}$$

$$UCL = (D4)(\bar{R})$$

$$LCL = (D3)(\bar{R})$$

Note 1: For the R chart, the lower control limit will very often be zero (there is no such thing as a negative control limit for this chart).

Note 2: A2, D4, and D3 are constants that have been especially developed for calculating control charts. Full sets of these tables are available from the American Society for Quality Control for all types of control charts; however, for the x-bar/R chart, the following table will get you started:

Factors for Control Limits in x-bar/R Control Charts

Subgroup Size	x-bar Factors	R Factors	
(n)	(A2)	(D3)	(D4)
2	1.880	0	3.267
3	1.023	0	2.575
4	0.729	0	2.282
5	0.577	0	2.115
6	0.483	0	2.004
7	0.419	0.076	1.924
8	0.373	0.136	1.864
9	0.337	0.184	1.816
10	0.308	0.223	1.777

The following is an example of an x-bar/R chart as well as how to put one together. Pressroom personnel at the Casey Printing Company wanted to monitor solid ink density variation on press, which is, of course, a classic case of a variable. The measurement unit is "density," and the measuring tool is the densitometer. The frequency of monitoring this variable for the purposes of the control chart might be one subgroup every 15 minutes. This interval could be determined by setting up the press and letting it run itself, without adjustment, until the variable under consideration began to drift significantly away from its starting point. The subgroup size is generally based on the number of samples required to get a quick snapshot of the process. In this case, let us assume three samples per subgroup. The next thing to do is to collect your samples for putting together the chart. As we start running, we take three press sheets (one subgroup) every 15 minutes, until we have twenty subgroups worth of data. The following is the data; x-bar and R for each subgroup as well as the grand average (x-double bar) and the average range (R-bar) have been calculated.

Subgroup	Reading	Reading	Reading	x-bar	R
1	1.56	1.46	1.48	1.50	0.10
2	1.50	1.42	1.48	1.47	0.08
3	1.56	1.50	1.52	1.53	0.06
4	1.47	1.41	1.37	1.42	0.10
5	1.47	1.49	1.45	1.47	0.04
6	1.55	1.52	1.44	1.50	0.11
7	1.50	1.45	1.48	1.48	0.05
8	1.48	1.32	1.40	1.40	0.16
9	1.52	1.59	1.63	1.58	0.11
10	1.59	1.47	1.38	1.48	0.21
11	1.40	1.54	1.46	1.47	0.14
12	1.51	1.48	1.48	1.49	0.03
13	1.50	1.56	1.50	1.52	0.06
14	1.52	1.52	1.48	1.51	0.04
15	1.46	1.45	1.46	1.46	0.01
16	1.41	1.45	1.34	1.40	0.11
17	1.44	1.47	1.43	1.45	0.04
18	1.44	1.34	1.50	1.43	0.16
19	1.54	1.48	1.44	1.49	0.10
20	1.52	1.46	1.47	1.48	0.06
			x-double bar	1.48	
			R-bar		0.09

The next thing to do is to calculate the control limits for the x-bar chart as follows:

$$\begin{aligned} CL &= \bar{\bar{x}} \\ &= 1.48 \end{aligned}$$

$$\begin{aligned} UCL &= \bar{\bar{x}} + (A2)(\bar{R}) \\ &= 1.48 + (1.023)(0.09) \\ &= 1.48 + 0.092 \\ &= 1.572 \end{aligned}$$

$$\begin{aligned} LCL &= \bar{\bar{x}} - (A2)(\bar{R}) \\ &= 1.48 - (1.023)(0.09) \\ &= 1.48 - 0.092 \\ &= 1.388 \end{aligned}$$

We also need to calculate the control limits for the R chart as follows:

$$\begin{aligned} CL &= \bar{R} \\ &= 0.09 \end{aligned}$$

$$\begin{aligned} UCL &= (D4)(\bar{R}) \\ &= (2.575)(0.09) \\ &= 0.232 \end{aligned}$$

$$\begin{aligned} LCL &= (D3)(\bar{R}) \\ &= (0)(0.09) \\ &= 0 \end{aligned}$$

Given these limits and data points, we are next ready to plot our chart and put the subgroup averages and ranges on their respective charts. This is shown at the top of the next page.

In interpreting these charts, it is important to look for the following patterns to determine if the process is in control:

1. All points within limits
2. General randomness
3. No trends or cycling patterns
4. No hugging of the center line
5. No more than four points in a row on one side of the center line
6. No more than one out of three points between the second and third standard deviation

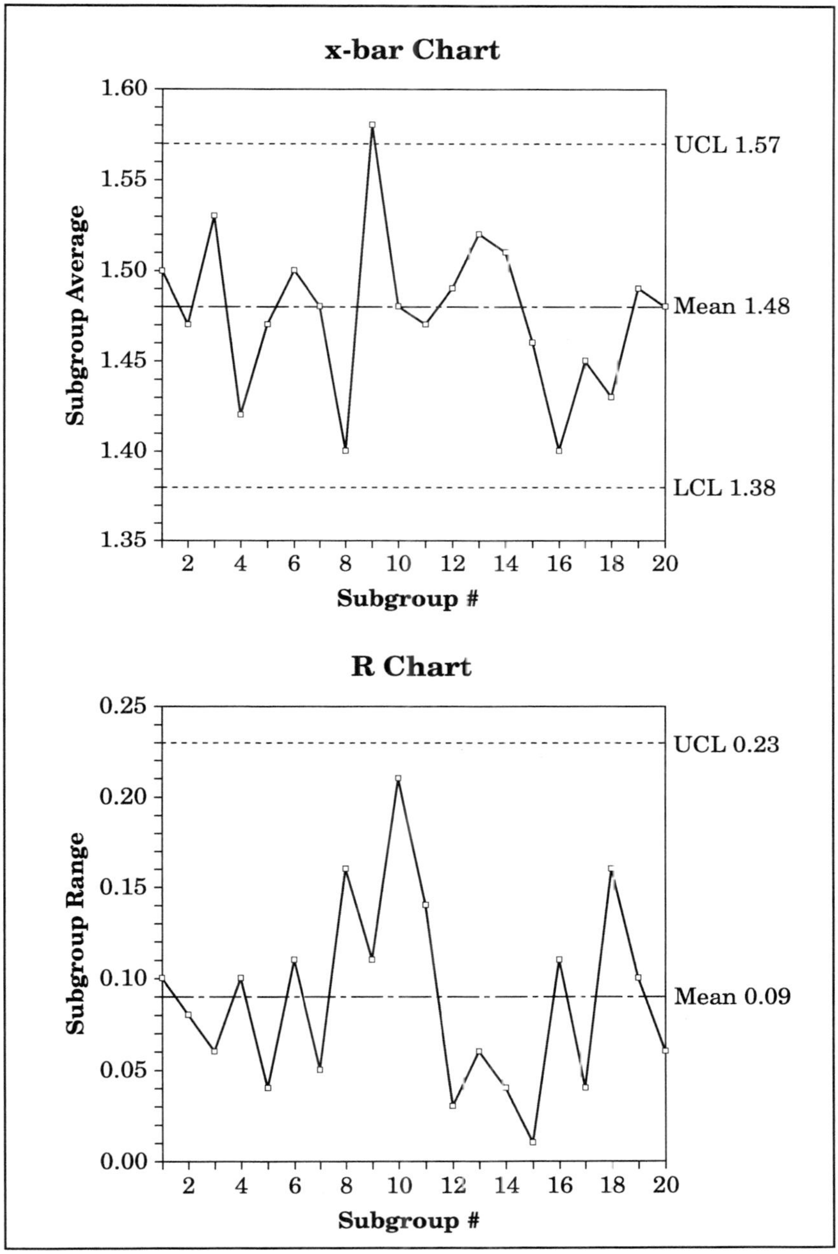

On the x-bar chart above, subgroup #9 is beyond a control limit. Since the corresponding point on the R chart isn't out of range, we can conclude that there probably wasn't an error of measurement causing that out-of-control signal. Therefore, the average density did jump at that point in time. This could be caused by any number of things, including the press operator filling the ink fountains too high, a more strongly pigmented ink being put in the press, and an adjustment to either the ink keys or the dampening rate. In any case, the press operator saw the warning signal and was able to make the appropriate adjustment so that the process is brought back into a state of control. This resolution and cause should be noted right on the chart for future reference.

On the R chart, while there aren't any points beyond a control limit, there does appear to be a recurring trend whereby the process variation increases over time, drop backs down, and begins to increase again. While the chart isn't definitive with regards to this pattern, the press operator should keep an eye on it and see if it recurs. The pattern seems to indicate that over a period of about 2–3 hours, there is a significant increase in the amount of variation in density. Maybe this time period is tied to another variable, such as ink temperature or fountain solution strength. The operator would have to analyze the situation more deeply to find out.

Monitoring Print Attributes

As stated earlier, there are two basic types of print attributes: defects and defectives. A defect is an individual item to be counted such as a hickey, scratch, or dog-eared page. A defective is any unit of product, such as a book, press sheet, or plate that contains one or more defects. The major differences between control charts for attributes and control charts for variables are:

1. Variables are monitored with two charts—one is for process average and one is for process variations; attributes are monitored with one chart, which is primarily a picture of the process average.
2. Variables charts require small subgroups; attributes charts require large subgroup sizes.
3. Attributes are, by definition, products that are already defective; variables can show you a problem before it causes the product to be defective.
4. Variables are based on the normal distribution; attributes are based on the binomial distribution.

The attribute control charts shown in the remainder of this chapter work with constant-size subgroups. This is *not* a prerequisite for attribute control charts. It is possible to work with variable-size subgroups; however, the control limits will change width with such charts. Thus, they are more easily worked with when one has access to a computer SPC package.

An example where this type of chart would be useful would be if one were charting plate makeovers as a percentage of plates made over time. The subgroup size (plates made) would probably differ (maybe quite significantly) from one sample to the next. Nonetheless, one could create a variable-size attribute chart from this data.

Counting Defectives: The p Chart

The ***p chart*** is used for looking at defectives as a proportion or fraction of the total sample. The formulas for constructing a p chart are:

$$CL = \bar{p}$$

$$UCL = \bar{p} + 3\sqrt{\frac{\bar{p}(1-\bar{p})}{n}}$$

$$LCL = \bar{p} - 3\sqrt{\frac{\bar{p}(1-\bar{p})}{n}}$$

where, "$\bar{p}$" (p-bar) is the average fraction defective of all subgroups and "n" is the size of each subgroup

The central line of the p chart is the average fraction defective (p-bar) for all subgroups. It is found by:

1. Calculating the fraction defective for each subgroup.
2. Adding up all fractions.
3. Dividing by the number of subgroups.

Control limits for the p chart are found by plugging p-bar (average fraction defective) into the above formulas.

In these formulas, "n" is the subgroup size. These control limits represent ±3 standard deviations from the mean and thus encompass 99.7% of the product under normal circumstances.

Example of the p Chart

In this example, we are looking at trim squareness from a three-knife trimmer. The printer has set up a template that is used for determining if a book is square or not. If it is not square, then it is categorized as defective and goes into the garbage bin. The printer is collecting data in subgroups of 100 books. These subgroups are collected at 30-minute intervals. With 25 subgroups, the accumulated data is as shown in the table on the next page.

Subgroup	Defective	Fraction
1	4	0.04
2	5	0.05
3	9	0.09
4	7	0.07
5	3	0.03
6	2	0.02
7	4	0.04
8	5	0.05
9	4	0.04
10	8	0.08
11	8	0.08
12	3	0.03
13	4	0.04
14	3	0.03
15	2	0.02
16	8	0.08
17	6	0.06
18	6	0.06
19	7	0.07
20	5	0.05
21	7	0.07
22	5	0.05
23	6	0.06
24	4	0.04
25	4	0.04
Totals	129	1.29

The next step is to calculate the central line and control limits for the data.

The central line is the total of the fractions divided by the number of subgroups:

$$\overline{p} = \frac{1.29}{25} = 0.0516$$

The upper and lower control limits for the data are calculated as follows:

$$= \bar{p} \pm 3\sqrt{\frac{\bar{p}(1-\bar{p})}{n}}$$

$$= 0.0516 \pm 3\sqrt{\frac{0.0516(1-0.0516)}{100}}$$

$$= 0.0516 \pm 3\sqrt{\frac{0.0516(0.9484)}{100}}$$

$$= 0.0516 \pm 3\sqrt{(0.0516)(0.009484)}$$

$$= 0.0516 \pm 3\sqrt{0.000489}$$

$$= 0.0516 \pm 3(0.022)$$

$$= 0.0516 \pm 0.066$$

Therefore the lines on the p chart will be:

Central line = 0.0516

Upper control limit = 0.0516 + 0.066 = 0.1176

Lower control limit = 0.0516 – 0.066 = –0.0144
(Lower control limits are set at zero anytime they are calculated as negative numbers.)

Next plot the data and interpret the chart. The chart is shown on page 170.

The interpretation of control charts for attributes is equivalent to interpreting a combined x-bar/R variables chart. The p chart shows both long-term variation (through its trend line) and short-term variation (through the distance between points). We are looking for the same types of trends, cycles, runs, out-of-control points, approaches to limits, and other nonrandom patterns as with variables charts.

In this case, there are no points out of limits. There seems to be the presence of two constellations of data: one centered around 0.035 and the other centered around 0.07. This could be the result of two different inspectors comparing the books to the template or it may indicate that the inspection process is using two different templates. There is also very extreme

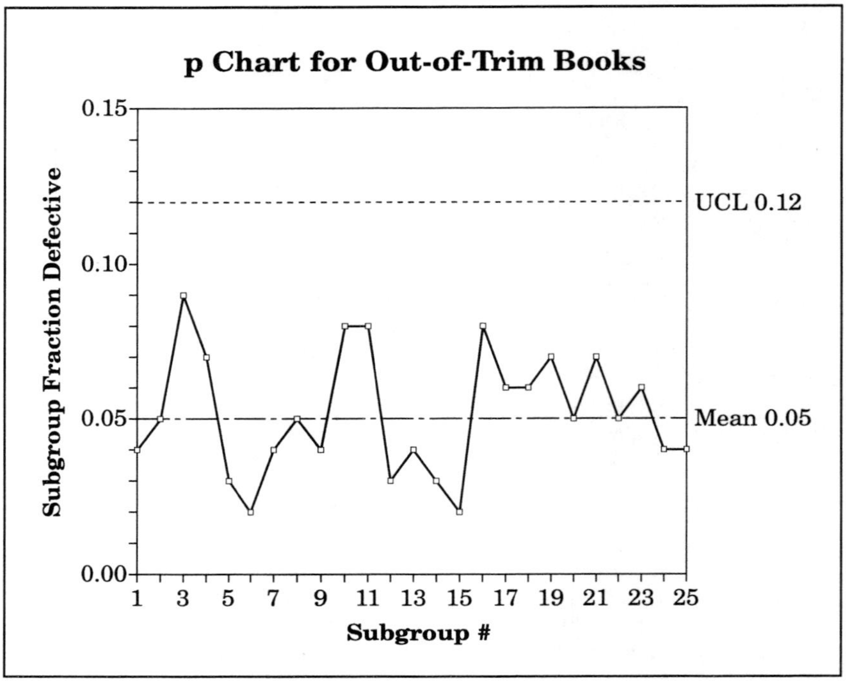

short-term variation exhibited on the chart. These extreme highs and lows could indicate that the operator is inexperienced or that the machine is wearing out (although machine wear often results in a trend, as well). In any case, this problem seems to decline toward the end of the chart. At the end of the chart, there seems to be a downward trend. This, of course, represents a process that is producing fewer defective products. All of these patterns should be investigated for assignable causes.

Before moving on to the next section, it should be noted that there is an alternative to the p chart. The ***np chart*** looks at defectives in absolute numbers rather than as a fraction of the whole. The effect is that you get the same chart only with the vertical axes shifted up. Other than that, everything else stays the same.

Counting Defects: The c Chart

As previously mentioned, it is sometimes preferable to look at the number of defects in a given sample rather than the number of defectives. This is particularly true when multiple defects can occur in a single unit. Examples include scratches per plate and hickeys per press sheet. For the purpose of counting defects, we use the ***c chart.***

The formulas for the c chart are as follows:

$$CL = \overline{c}$$

$$UCL = \overline{c} + 3\sqrt{\overline{c}}$$

$$LCL = \overline{c} - 3\sqrt{\overline{c}}$$

In these formulas "$\overline{c}$" (c-bar) is equivalent to the total number of defects divided by the number of subgroups. The central line is, therefore, the average number of defects per subgroup. The control limits are again set at ±3 standard deviations from the central line. Interpretation is the same as with the p chart.

Example of the c Chart

In this example, we are going to look at the number of hickeys in samples of 50 press sheets. The samples are collected at intervals of 30 minutes over a 12.5-hour period of time for a total of 25 samples. Let us look at the data:

Sample	# Hickeys
1	8
2	9
3	5
4	8
5	5
6	9
7	9
8	7
9	8
10	7
11	6
12	4
13	7
14	6
15	9
16	6
17	4
18	11
19	7
20	20
21	18
22	6
23	9
24	10
25	5
Total	203

From this we can now calculate the central line and the control limits.

The central line, c-bar, is equal to the total number of hickeys divided by the number of samples:

$$\bar{c} = \frac{203}{25} = 8.12$$

The upper and lower control limits are equal to c-bar ±3 times the square root of c-bar:

$$= \bar{c} \pm 3\sqrt{\bar{c}}$$

$$= 8.12 \pm 3\sqrt{8.12}$$

$$= 8.12 \pm 3(2.85)$$

$$= 8.12 \pm 8.55$$

Therefore the limits on the chart are:

Central line = 8.12

Upper control limit = 8.12 + 8.55 = 16.67

Lower control limit = 8.12 – 8.55 = –0.43
(Note: LCL is zero if negative.)

The control chart for c from this data looks like:

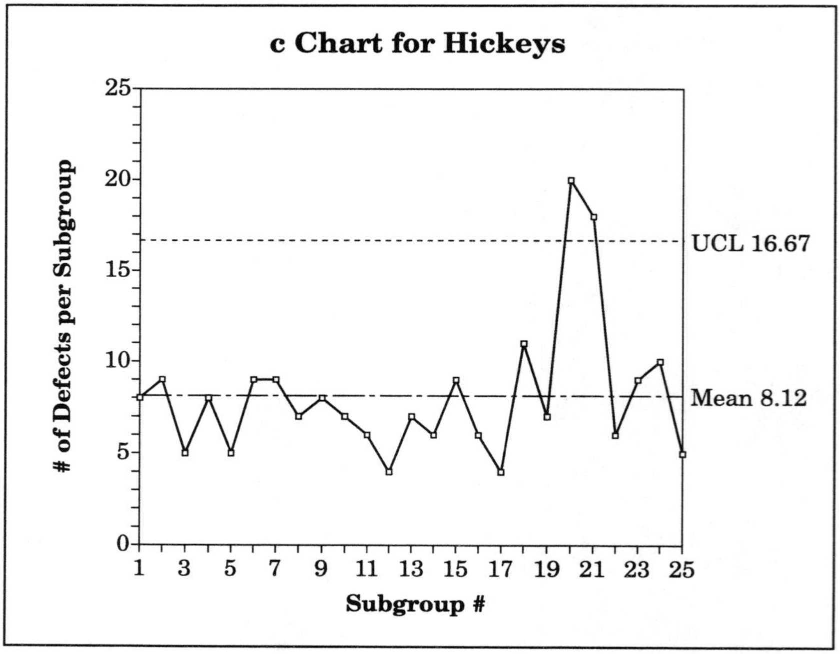

On this chart, sample numbers 20 and 21 are beyond the upper control limit. Something has changed the process so that it has produced noticeably more defects. On investigation, it was found that the operator opened up and started using a can of ink that had been on the shelf for a while. It had greater-than-average ink skin, which resulted in an increase in hickeys.

With all control charts, once an assignable cause has been detected and removed from the system, the control limits should be revised without the out-of-control points. By so doing, the control chart will continue to reflect only random variation. Control limits should generally not be revised unless one is certain that there is an improved state.

If we recalculate the limits without points 20 and 21, we would attain the following:

Total hickeys = 165

C-bar = 165/23 = 7.17

UCL = $7.17 + 3\sqrt{7.17} = 15.2$

LCL = $7.17 - 3\sqrt{7.17} = -0.86$ (zero for the chart)

The revised chart would look like this:

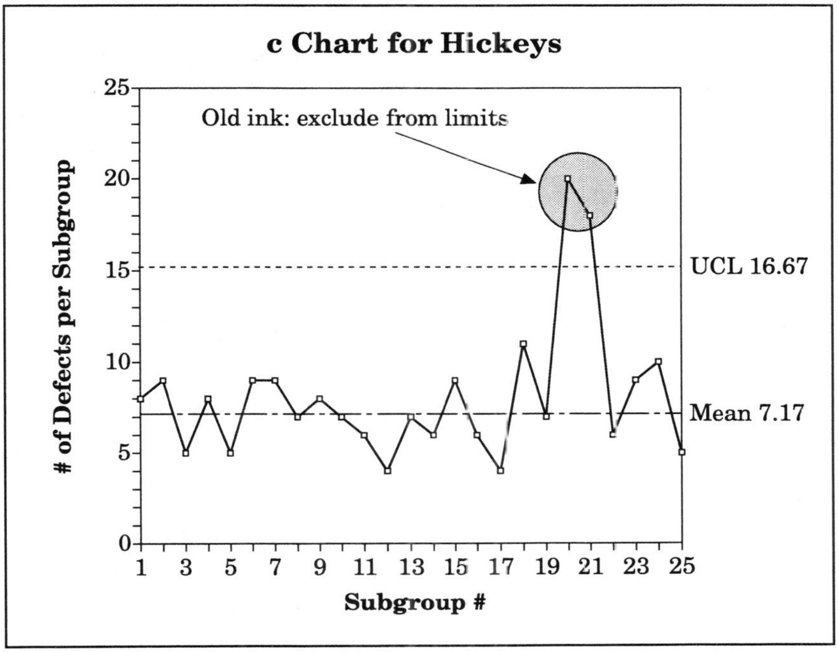

This procedure, whereby points are isolated and removed from control limits, is standard for all types of control charts. It is also much easier to do with a computer than manually.

Another variation of the control chart for counting defects is called the ***u chart.*** It is similar to the c chart, except that it breaks the defects down by unit rather than by sample. As with the np chart mentioned above, the effect is merely to shift the axes up or down.

When attempting to use a control chart, it is often helpful to have a standard form to work from. The form on page 175 is an example of a form for attributes charting.

Summary

Practical applications of control chart techniques have been dealt with in this chapter. Control charts for variables are used to monitor characteristics that can be measured. Sometimes, it is either impossible or uneconomical to measure quality characteristics. This then becomes the domain of control charts for attributes, where characteristics are judged as good-bad, go/no-go, or present-absent. These charts are less useful than variables control charts for improving the process and preventing defective product from being produced. Nevertheless, they have an important place, especially where the product is already being inspected. If inspection is happening anyway, using the data in a control chart for attributes can lead to substantial quality gains. Whether monitoring an attribute or a variable, the control chart is a useful diagnostic tool for an operator or a manager to evaluate the performance of a process over time. It is also an essential step toward quantifying process capability, which can only first be truly evaluated when a process is known to be in control.

Casey Press Attribute Control Chart Form

Chart Type p ☐ c ☐	Attribute Being Charted	Sample Size	Frequency

Central Line = UCL = LCL =

Sample (n)	
Number Defective	
Proportion Defective	

Formulas

p Chart		c Chart	
Central Line	3-Sigma Limits	Central Line	3-Sigma Limits
$\overline{p}$	$\overline{p} \pm 3\sqrt{\overline{p}(1-\overline{p})/n}$	$\overline{c}$	$\overline{c} \pm 3\sqrt{\overline{c}}$

Action on Assignable Causes

Make note of any change in materials, methods, machines, people, or environment that could lead to an assignable change in the process.

Date	Time	Change	Action Taken (if any)

8 Design and Analysis of Experiments

Comparison of Design of Experiments to Statistical Process Control

Design of experiments (DOE) is a whole discipline of techniques dealing with the systematic analysis of input and output process variables. DOE is not a component of statistical process control (SPC) but rather a separate component of the total quality system. To understand how DOE and SPC differ, examine the following chart:

Statistical Process Control

- Involves the acquisition, analysis, and use of information already being produced at random within the process.
- Passive data acquisition. The observer waits for something to happen and then attempts to discover what happened.

Design of Experiments

- Key input variables (factors) are manipulated in a planned manner, and the output (response) is monitored for resultant changes.
- Active data acquisition. The experimenter manipulates variables to examine their effects on results.

To demonstrate how each of these two methods might be applied in the printing industry, consider the example on the next page that looks at register variation.

Both DOE and SPC are necessary components of the total quality system, each with its own areas of application. Knowing the value of each and when they should be used results in a more effective overall system. A scatter diagram (with correlation coefficient determination), as described in Chapter 5, is a simple example of DOE, where only one variable is manipulated to determine the effect on one other variable.

Statistical Process Control

- A control chart is established to monitor register variation.
- Detection of abnormal (assignable) causes is noted.
- Adjustments to the process are made based on discovered causes. Otherwise, the process is left alone.

Design of Experiments

- Key input variables (factors) are manipulated (i.e., press speed, web tension, etc.) in a deliberate and planned manner. The output is monitored for changes.
- An analysis is made of the relationships between input and output variables.
- The process is "forced" to yield information about itself, in order that improvement can take place.

Learning and the Scientific Method

In any type of investigation, attempts are made to learn something about a particular area of concern. In order for learning to take place, two things must happen: (1) a "critical event" and (2) a "perceptive observer."

Reduced to simplest terms, what this means is that in order to learn anything, something must happen and we must be ready to see it. While this may seem simple enough, carrying it out is not so easy, since "critical events" tend to happen when least expected (at random), and sensitive means of detecting these events are not always in place at the right time. Additionally, the probability of both occurring simultaneously is even lower. As a result, learning is a slow, tedious process. In fact, many companies have gone bankrupt for lack of efficient learning processes.

The goal of the scientific method is to find a way of bringing these two items together at the same time by using two fundamental tools:

- **Informed observation.** This is the process of knowing what to look for so that when the "critical event" occurs, it will not go unnoticed. This is the realm of SPC.
- **Directed experimentation.** The process of forcing something to happen. This is the domain of DOE.

History and Fundamentals of DOE

The need to experiment seems to be a human condition. This need has probably been around since the beginning of time. From the cave people experimenting with different methods for sharpening spears to make them more efficient, through such famous experimenters as Mendel or Galileo, experimentation has yielded much information about nature and the way things work. However, it was not until the 1920s that the science of experimentation matured as a means of calculated progress in all facets of life.

Probably the greatest contributor to the science of experimentation was an Englishman by the name of R. A. Fisher. Fisher, working in the 1920s, was challenged with improving the yield of crops such as barley and potatoes under the adverse conditions found in the British Isles. Of course, the experiments could have been conducted in greenhouses, isolated from natural conditions, but this might have produced results that could not be duplicated in the real world. Fisher decided, therefore, to conduct the experiments in the field, where he could be certain that the results would be repeatable.

The primary obstacle was how to tell if his experiments were having any effects, since the real world is so full of random events.

Fisher developed a process of isolating experimental effects (called "signals") from random events (called "noise"). Fisher developed several unique ways of enhancing the signal, which made his experiments more efficient. Three of the more important techniques that Fisher used were:

- **Blocking.** The systematic isolation of differences between various experimental conditions. The term comes from the "blocks" of land Fisher had to work with.
- **Randomization.** The assigning of various experimental combinations (called "treatments") within a single block in random order.
- **Replication.** The repetition of experiments until the signal becomes discernible, allowing the experimenter to detect even small signals.

It is natural for things to happen at random. This is true in printing as well as in Fisher's plots of land. So Fisher did the only thing remaining to be done: he conducted his experiments repeatedly until he could hear the "signal."

Various tests are available (chi-square, t, and f) which allow us to determine the degree of confidence detected in a signal. These will be referred to again in the section covering the evolutionary operation (EVOP) experimental design.

Isolating the Signal in the Noise

In the following diagram, you can see the effects of replication, randomization, and blocking, as each serves to isolate the experimental signal and remove the natural process noise.

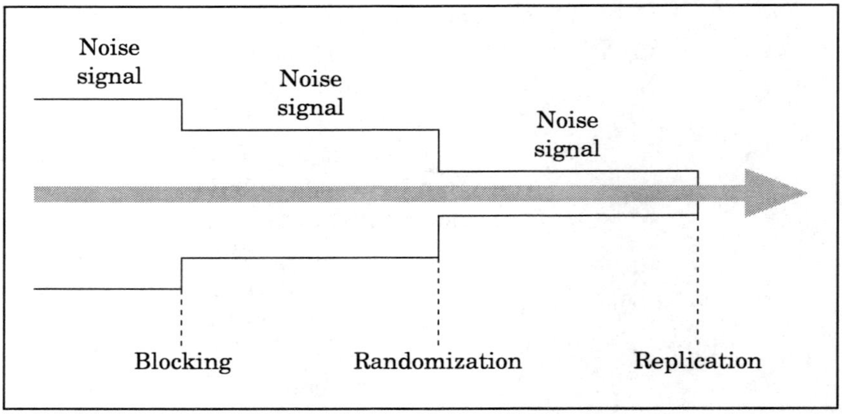

To demonstrate the application of these three techniques to the printing industry, we will use an experiment that looks at the relationships of different types of offset blankets, different ink densities, and different press speeds on the effect of dot gain. Our goal is to determine the best combination of factors for producing a product with the lowest dot gain.

Using each of Fisher's techniques, we would take the following approach:

Blocking. If the experiment is being run on two different presses, it is a possibility that each press has its own characteristic dot gain. Other sources of block variability include:

- Day-to-day differences
- Shift-to-shift differences
- Run-to-run differences

Blocking allows the isolation of these differences in order that the experimenter can concentrate on the other factors that are of interest.

Randomization. Are there unexpected trends or patterns of variability on a single press? Randomization allows us to run random experiments calling for different density/blanket/speed combinations. Running these experimental trials in random order will force results that are due only to the treatments and the random variation, not the periodic ups and downs of the press.

Replication. With batch-to-batch differences eliminated by blocking, and trends and cycles eliminated by randomization, the only "noise" hiding our "signal" is random variation. Thus, we would conduct the experiment repeatedly until the "signal" becomes discernible.

One Variable at a Time (OVAT) vs. DOE

Fisher had another revolutionary idea for making his experiments more efficient. Traditionally, experiments had been conducted by moving "one variable at a time." Fisher's idea was to move all of the variables simultaneously, which he called a "factorial design." By moving all variables simultaneously, it is possible to detect not only individual effects, but also the effects of different combinations of factors. Such combination effects are called "interaction effects."

To use an example from the printing industry that can illustrate the differences between OVAT and DOE, suppose we wanted to study the effects of temperature, chemical strength, and length of agitation in the processing of plates. The following diagram shows how the experiment would be approached from each of the two methods.

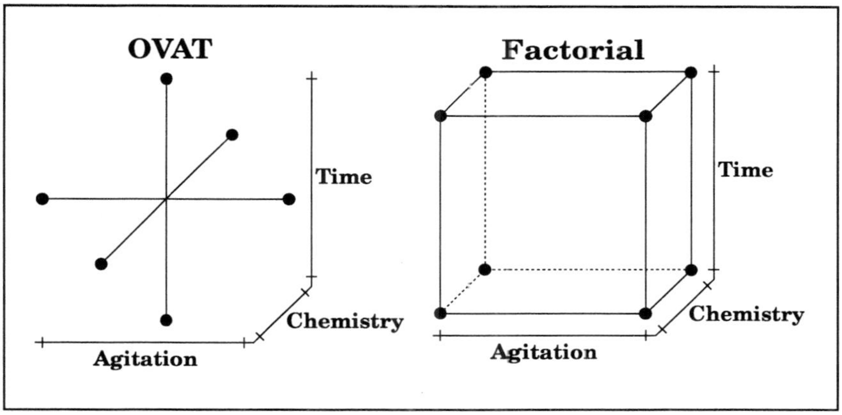

OVAT vs. DOE. The "traditional" OVAT design, on the left, shows how each variable is moved, one at a time, some distance. A response is measured and then the variable is returned to its original setting. This design will provide all the information available for each variable by itself. After running the OVAT, it will be determined to what extent agitation, chemical strength, and temperature individually effect the output of the response. The question remains, "Are there any interaction effects?"

To determine if there is interaction among the variables, we must apply a factorial design like the one on the right. For example, is the combined increase of chemical strength and temperature greater than the individual effects? This approach yields much greater information than the OVAT method. In general, each type of design will yield the following pieces of information:

Three-Variable OVAT	**Three-Variable Factorial**
Six individual effects	Six individual effects Three two-factor interaction effects One three-factor interaction effect

Let us now take a specific printing example to demonstrate how a three-factor, two-level experimental design (32) would be set up. In this example the effects on dot gain caused by the following three input variables are investigated: two brands of blanket, various solid ink densities, and various press speeds. This is called a two-level design, because each run looks at only two levels for each factor.

The parameters we will use for the first run of this experiment are as follows:

Factor	**Low Level**	**High Level**
Blankets	Brand "A"	Brand "B"
Density	1.0*	1.15*
Speed	35,000 iph*	45,000 iph*

*These numbers have been chosen arbitrarily. In reality, to choose the levels for a designed experiment, it is necessary to have good knowledge of the process. The range over which the factors would be varied will depend on the experimental objectives.

Steps in Designing the Experiment

Step 1. Establish a plan for implementation of the experiment. First of all, it must be determined if blocking will be necessary. Let us assume that this experiment will be run on two different presses. Since press-to-press differences might exist, it will be necessary to block this design. For a 32 (three-factor, two-level) design such as this, the easiest way of planning the blocking is to draw a diagram, such as the one that follows:

Blocking Diagram

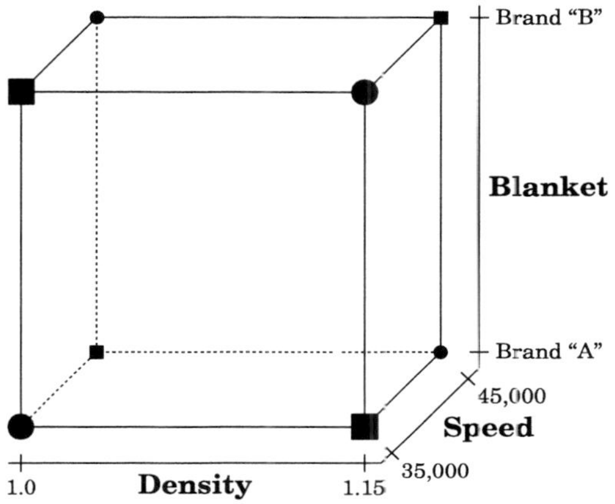

The boxes in the diagram represent the runs that will take place on press "a," and the circles are for the runs on press "b." Notice that there are two runs for each press on every side of the cube. This balancing act will systematically eliminate any effects that result from differences between the two presses. How to double-check the blocking's effectiveness is explained later.

Step 2. Put together the remainder of the plan. The following table shows how the experimental plan is established:

Run	Density	Speed	Blanket	Press	Trial 1
1	1.0	35,000	A	b	b3
2	1.15	35,000	A	a	a3
3	1.0	45,000	A	a	a1
4	1.15	45,000	A	b	b2
5	1.0	35,000	B	a	a4
6	1.15	35,000	B	b	b1
7	1.0	45,000	B	b	b4
8	1.15	45,000	B	a	a2

Although this plan, at first glance, might appear confusing, it is really straightforward in structure. The method for determining the various factor combinations found in the second, third, and fourth columns is key to the design. Notice that with the first factor, density, the level is alternated between each run. For the second factor, speed, the level is alternated every second run. Finally, for the last factor, blanket, the level is alternated every fourth run. If there was a fourth factor, it would be alternated every eighth run (of course, such a design would require sixteen, rather than just eight, runs to complete). By combining the factors in this way it is assured that all combinations of factors are tested throughout the experiment.

Step 3. Determine which combinations will be run on which press. To make this determination, we go back to the original block design. All that is required is to compare the different combinations in the table with the coordinates in the block design. For example, looking up the combination of density of 1.15, press speed of 35,000, and blanket type "B" in the block design shows a circle. This corresponds to press "b." Thus, run #6 will take place on press "b." This procedure is followed for each set of treatment combinations, until the column "Press" is complete.

Step 4. Allocating the runs at random. There are four runs per press (block). Each set of four is randomized and run in the order determined. For this task, any random numbers table will do. By the process of randomizing runs, any systemic noise within a single block is reduced.

Notice that this column is titled "Trial 1." One trial is unlikely to yield statistically significant results. Therefore, the design must be run through several trials with analysis taking place after each trial. In running successive trials, the only thing that must change is the random order of the runs within blocks. Everything else stays the same. The running of successive trials is the process of replication. Let us now look at how we can analyze the results from this experiment.

Analysis of Experiment

After running the design through one cycle, analysis of the results with regard to individual and interaction effects can begin. The diagram at the top of page 185 shows the results from the first cycle (the numbers given are for dot gain in the midtone area).

For example, the results from cycle #1 of the combination of density of 1.15, speed of 35,000, and blanket type "B" was a midtone dot gain of 30%. A quick look at the rest of the

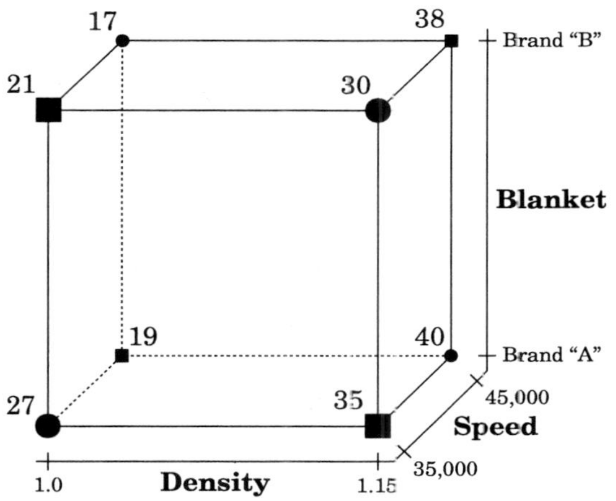

numbers shows an overall range from a low of 17% to a high of 40%. This raises the following question, "Which factors or combinations of factors caused such a wide range of results?" To answer this question, we must analyze the individual effects and the interaction effects.

First, analyze the effects of the individual factors.

For the ***density effect,*** compare the dot gains at the left side (where density was 1.0) to the dot gains on the right side of the design (where the density was 1.15):

Density Effect

Density = 1.0	Density = 1.15	Difference
17	38	21
21	30	9
19	40	21
27	35	8
	Average Difference	14.75

From this we can conclude that the effect of changing density from a lower to a higher level was an average increase of 14.75% dot gain.

For the ***blanket effect,*** compare the dot gains on the bottom of the cube for blanket "A" with those at the top of the cube for blanket "B":

Blanket Effect

Blanket "A"	Blanket "B"	Difference
27	21	–6
19	17	–2
35	30	–5
40	38	–2
	Average Difference	–3.75

From these calculations, it can be concluded that the average effect of going from blanket "A" to blanket "B" is a net dot gain reduction of 3.75%.

For ***press speed effect,*** compare the dot gains at the front of the cube (speed of 35,000) with those at the back of the cube (speed of 45,000):

Press Speed Effect

Low Speed	High Speed	Difference
21	17	–4
30	38	8
27	19	–8
35	40	5
	Average Difference	0.25

From this, it can be concluded that the individual effect of changing press speeds is a negligible increase of 0.25% dot gain.

If the analysis ceased at this point, it might be tempting to believe that only density is a real factor in minimizing dot gain. This would be a mistake. To get a complete picture of the results of the experiment, the interaction effects of these three factors must be analyzed.

Interaction effect of density and speed. The density-speed interaction is essentially a comparison of the density effects at the front of the cube (where speed in 35,000) to the density effects at the back of the cube (where speed is 45,000). This is calculated as follows:

Density effect at low speed: [(35 – 27) + (30 – 21)] ÷ 2 = 8.5

Density effect at high speed: [(40 – 19) + (38 – 17)] ÷ 2 = 21

Density-speed interaction effect: (21 – 8.5) ÷ 2 = 6.25

This means that, although speed by itself has practically no effect, the joint effect of speed and density is an increase in dot gain of 6.25%. Thus, varying density has a much greater impact at higher speeds than at lower speeds. It also means that speed ***is*** relevant, even though its individual effect was marginal.

Interaction effect of speed and blanket. The speed-blanket interaction is a comparison of the speed effects at the bottom of the cube (with blanket "A") to the speed effects at the top of the cube (with blanket "B"). This is calculated as follows:

Speed effect with blanket "A": [(40 – 35) + (19 – 27)] ÷ 2 = –1.5

Speed effect with blanket "B": [(38 – 30) + (17 – 21)] ÷ 2 = 2

Speed-blanket interaction effect: [2 – (–1.5)] ÷ 2 = 1.75

Thus, the speed-blanket interaction is negligible.

Interaction effect of density and blanket. The density-blanket interaction is a comparison of the density effects at the bottom of the cube (with blanket "A") to the density effects at the top of the cube (with blanket "B"). This is calculated as follows:

Density effect with blanket "A": [(35 – 27) + (40 – 19)] ÷ 2 = 14.5

Density effect with blanket "B": [(30 – 21) + (38 – 17)] ÷ 2 = 15

Density-blanket interaction effect: (15 – 14.5) ÷ 2 = 0.25

This is also a negligible interaction. All of these effects can be summarized as follows:

Individual Effects		**Interaction Effects**	
Density	14.75	Density-speed	6.25
Blanket	–3.75	Speed-blanket	1.75
Speed	0.25	Density-blanket	0.25

The conclusions are as follows:

- Density by itself has the greatest individual effect on dot gain.
- Blanket type "B" showed an average dot gain 3.75% less than blanket type "A."
- Speed, by itself, seems irrelevant.
- The joint effect of density and speed shows that the impact of density on dot gain is much greater (6.25%) at the higher level than at the lower level. Therefore, speed is relevant.
- The other joint effects seem minimal.

To test the significance of these conclusions would require the use of statistical significance tests (which are straightforward but beyond the scope of this chapter).

As previously noted, a skeptical reader may want to test the blocking effect. To do so, add some quantity (for example 5%) to the responses generated by one of the two presses. This simulates the effect of systemic, press-to-press differences. The reader is then invited to re-analyze the effects. It will be discovered that the balanced arrangement of runs allows for any differences. Thus, the effects will remain unchanged.

EVOP and Response Surfaces: A Model for Process Optimization

Many different experimental designs can be derived from the basic two-level factorial model just described. One model, which, in the opinion of these authors, can be of particular use to printers, is called ***evolutionary operation*** (EVOP). On the following pages, we discuss EVOP and the underlying concept of the response surface. Following this, an example of the EVOP design is presented and analyzed.

EVOP, along with many other experimental designs, is based on the concept of the response surface. Simply stated, the response surface is a graphic illustration of the manner in which a factor responds to varying levels of process inputs.

EVOP was originally designed by the prominent statistician George E. P. Box in the late 1950s. Box based the design on the Theory of Evolution, in which nature takes advantage of beneficial random events over long periods of time to produce a better product. The only difference between nature and the printing industry is the product: one produces life, the other produces printed material.

The EVOP model soon gained wide acceptance in the chemical industries, which were often concerned with achieving maximum yields from batch processes. In these industries, such variables as temperature, chemical strength, and agitation time were manipulated and the batch yield was monitored as a response. Essentially, any time when a process is established and in production, the EVOP design can be applied.

The following illustration should help the reader to conceptualize what is meant by this:

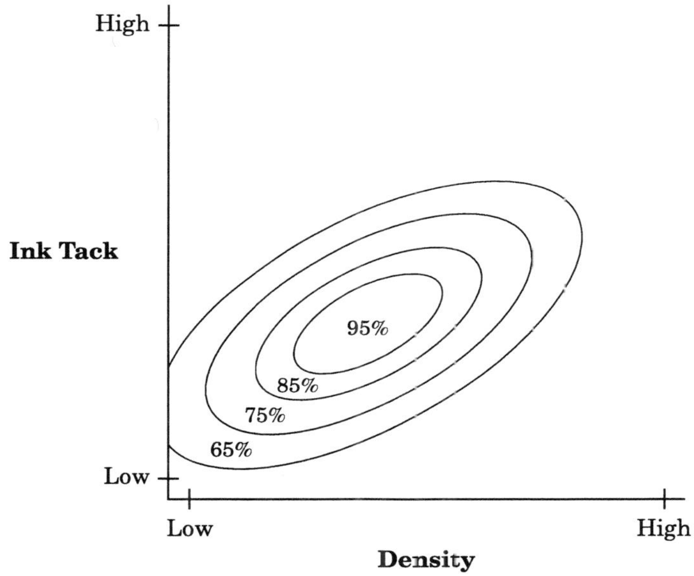

This illustration shows a theoretical response surface for percent wet ink trapping as a function of two process inputs: ink density and ink tack. The elliptical patterns in the graph represent the trapping response to varying levels of the inputs. These patterns can be interpreted in the same way as a topographic map. The outermost ellipse represents the lowest trapping. Movement towards the center shows increased response.

In three dimensions, this would appear as a mountain, which has a central peak at elevation 95%. As the illustration shows, very little change exists in the level of the response when only one variable is moved, but if both are moved, an optimal point can be quickly reached. Mathematically, this is equivalent to the interaction effect of the two input variables.

The goal of the EVOP design is to move up the response mountain and allow the experimenter to reach optimal conditions.

Most experimental designs are revolutionary in the sense that large changes are made in the input variables in order to force the process to yield information about itself. However, a major problem exists with the application of this type of revolutionary experiment: the experimental changes are likely to produce at least some out-of-specification product. When we experiment on a process that is already producing an acceptable product, the prime directive is to keep manufacturing acceptable product. Otherwise, waste and monetary loss will result. This conflict is resolved by making only very small changes to the process inputs, and the EVOP design provides the framework for this type of experiment, without the losses.

A few facets of the printing process that EVOP can help to optimize include the following:

- Trim and folder variation
- Density variation
- Dot gain and trapping levels
- Rub and scuff resistance

EVOP itself is a multiple-factor, two-level design, such as those already discussed, which moves through a series of phases. The EVOP crawls up the slope of the response mountain, attaining more optimal conditions with each phase. Additionally, the design uses a center point, which corresponds to current operating levels for the inputs. The center point is used to analyze the slope of the response curve and tell if an optimal condition has been reached.

Since the EVOP uses only very small changes in the input variables, running each phase through several replications is usually necessary. Each replication within a phase is called a cycle. With each new cycle, more of the noise associated with the random variations of the process is isolated, and detecting the signal becomes easier. Only when the signal is clear enough (it must pass a statistical test for significance) does the experiment move on to a new phase.

The diagram on page 191 shows how the EVOP climbs the response surface to optimal conditions.

In this diagram, the EVOP went through four phases to reach optimal conditions. It should be noted that this is a very slow procedure. Since the EVOP only deals with incre-

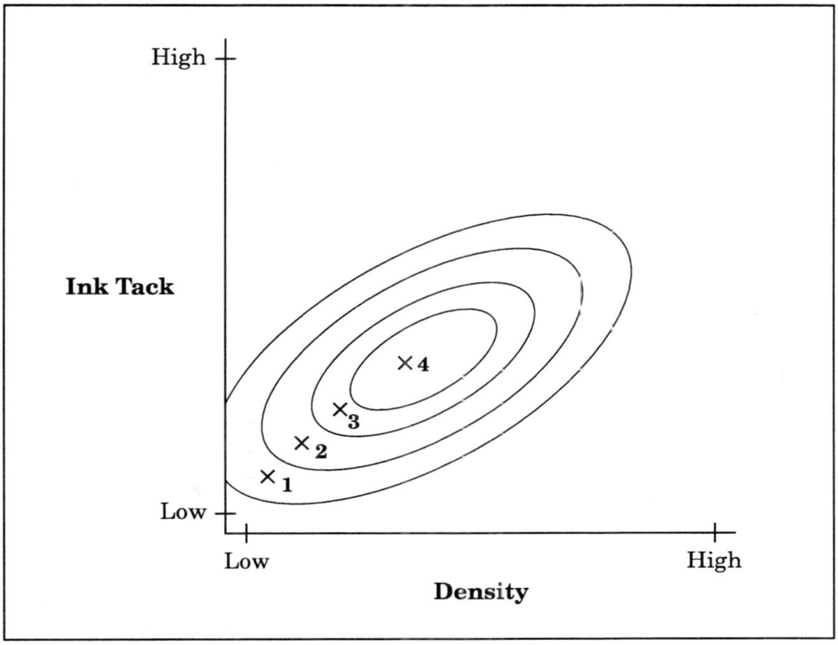

mental changes, the many cycles necessary to gather reliable data takes time and patience. The typical EVOP life cycle may be as long as several weeks; therefore, only those who are seriously committed to process improvement should use the design. EVOP is not a quick fix, but using it to produce a higher-quality product can provide printers with a potent market advantage.

When Box originally designed the EVOP, he had in mind that production personnel would be responsible for conducting the experiment. To make this burden a little lighter, he had the wisdom to put the EVOP into a standard format, with basic calculations and straightforward instructions for analysis.

Steps for Running the EVOP

The following guidelines should be used when developing and implementing an EVOP experimental design:

1. Working as a team, decide which inputs and responses are to be examined.
2. Decide what the acceptable ranges are over which the input variables can be moved.
3. Complete the EVOP worksheet on page 192, filling in first the phase number, conditions under investigation, levels for the phase, and which run (cycle) of that phase the experiment is in.

Chart 1

Two-Variable EVOP Worksheet

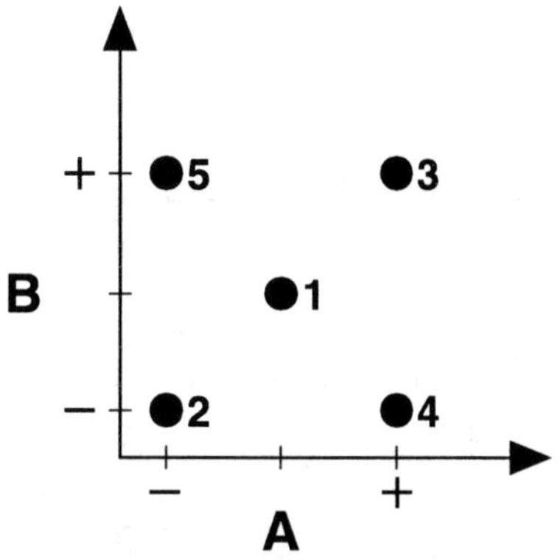

Project ______________________

Phase ______________________

Cycle n = ______________________

Response ______________________

Date ______________________

Calculation of Averages

Operating Conditions	1	2	3	4	5
(i) Previous Cycle Sum					
(ii) Previous Cycle Average					
(iii) New Observations					
(iv) Difference (ii – iii)					
(v) New Sums					
(vi) New Averages: y_1					

Calculation of Effects

"A" effect = $1/2(y_3 + y_4 - y_2 - y_5)$ =

"B" effect = $1/2(y_3 + y_5 - y_2 - y_4)$ =

"AB" effect = $1/2(y_2 + y_3 - y_4 - y_5)$ =

CIM effect = $1/5(y_2 + y_3 + y_4 + y_5 - 4y_1)$ =

Calculation of Standard Deviation

Previous $\sum s$ =

Previous average s =

New s = (range)($f_{k,n}$) =

New $\sum s$ =

New average s =

Calculation of Error Limits

New average = $(2/\sqrt{n})(s)$ =

New effects = $(2/\sqrt{n})(s)$ =

For CIM = $(1.78/\sqrt{n})(s)$ =

n	$f_{k,n}$
2	0.3
3	0.35
4	0.37
5	0.38
6	0.39
7	0.40
8	0.40
9	0.40
10	0.41
11	0.41
12	0.41
13	0.41
14	0.41
15	0.42

4. Run the first cycle of the first phase: *treatment combinations should be run in random order.*
5. Fill in the observations from the first cycle.
6. Run the second cycle of the first phase of the EVOP.
7. Calculate the new sums and new average observations, using the observations from the first and second cycles of that phase.
8. Calculate the effects according to the formulas found on the EVOP worksheet. (These are similar to those that were calculated earlier in the chapter. However, the CIM—change in mean—effect is new to us. The CIM effect is an average that compares the center point to the four corner points. If the CIM is negative, the center is higher than the corners. This would indicate that the top of the response mountain has been reached with optimal conditions. If the CIM is positive, the center point is lower than the surrounding corners. This indicates that the experiment is located at a minimum. The CIM is used to indicate when to discontinue the experiment.)
9. Calculate the standard deviation by first subtracting the new observations from the old ones. The absolute range between the highest and lowest of these differences is multiplied by the factor "f," which is used to convert a range to a standard deviation statistic.
10. Calculate error limits, which have been set (by our choice of "f") at the 95% level. Use the new average standard deviation in the formulas for these calculations.
11. Compare effects with limits. If any of the effects are found to be in excess of their corresponding limits, move the EVOP to a new phase level. If the CIM effect is negative and in excess of its limit, discontinue the EVOP.

EVOP Example

The evolutionary operation will be illustrated with a real-life situation, which deals with an issue of concern to many printers, namely how to get the longest plate run-length achievable.

One reason why litho plates wear out is that the image wears out. Many printers try to "harden" the image to reduce wearing out. Inputs that can affect the hardness of an image include exposure and processing variables. This example examines two input variables: exposure (as measured in units of light) and processing time (as measured in seconds). Many other variables play a role in plate wear (ink grind, fountain solution pH, pressure settings, etc.), and the EVOP, like other factorial designs, can handle many factors at once. However, for the sake of simplification, the EVOP example looks at only the two cited above.

It should also be noted at this point that tradeoffs exist, which should be taken into account. For example, increasing exposure hardens the image, but it may also lead to excessive dot gain. Tradeoffs like this can be monitored as additional responses, giving the experimenter information on acceptable ranges for input variables. When one considers that the EVOP looks at many variables and their interactions, and gives the user information pertaining to tradeoffs, the usefulness of the design becomes readily apparent. In this regard, EVOP is a very realistic design.

Step 1. Select proper experimental ranges. The following settings and ranges were determined through tests and prior experience for the first EVOP phase:

Current setting for exposure = 35 units
Safe range* = ±2 units

Current processing setting = 95 seconds
Safe range* = ±2 seconds

Step 2. Begin running the experiment. Having selected proper ranges for the first phase, proceed to the first and second cycles of the EVOP. Remember to be patient with this process; it will take time!

Run phase one, cycle one. The results are recorded on the EVOP worksheet, as shown in chart 2 on page 195.

The range of observations went from a low of 830,000 impressions per plate to a high of 950,000 impressions per plate. Although nothing can be conclusively stated from the first cycle (for lack of data), the astute experimenter will note that the setting of 37 units and 97 seconds yielded a full 100,000 impressions more than the current setting of 35 units and 95 seconds. The second cycle will yield more information concerning this phenomenon.

---

*Safe ranges must be determined by testing and prior experience. Safe ranges are defined as the selection of settings that will not lead to substandard quality, yet are wide enough to induce statistically noticeable differences in the response. With exposure, it is important to take into consideration items such as dot gain and dot sharpening. With processing it is important to look at characteristics like adequate washout to delineate acceptable experimental ranges.

Chart 2

Two-Variable EVOP Worksheet

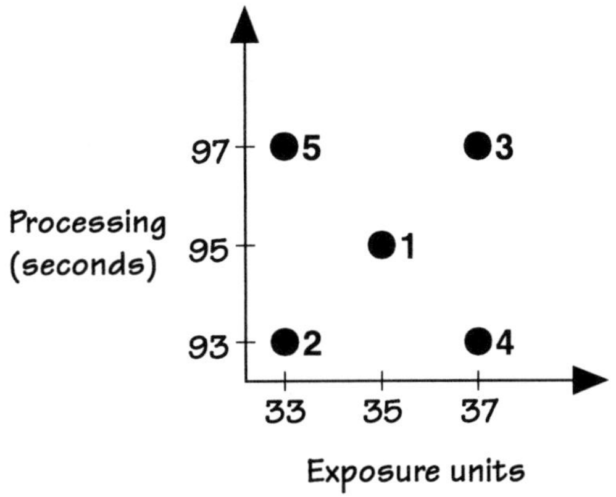

Project: Example

Phase: 1

Cycle n = 1

Response: run length

Date: 4-5-1995

Calculation of Averages					
Operating Conditions	1	2	3	4	5
(i) Previous Cycle Sum					
(ii) Previous Cycle Average					
(iii) New Observations	90	85	95	85	83
(iv) Difference (ii – iii)					
(v) New Sums	90	85	95	85	83
(vi) New Averages: y_1	90	85	95	85	83

Calculation of Effects

"A" effect = $1/2(y_3 + y_4 - y_2 - y_5)$ =

"B" effect = $1/2(y_3 + y_5 - y_2 - y_4)$ =

"AB" effect = $1/2(y_2 + y_3 - y_4 - y_5)$ =

CIM effect = $1/5(y_2 + y_3 + y_4 + y_5 - 4y_1)$ =

Calculation of Standard Deviation

Previous $\sum s$ =

Previous average s =

New s = (range)($f_{k,n}$) =

New $\sum s$ =

New average s =

Calculation of Error Limits

New average = $(2/\sqrt{n})(s)$ =

New effects = $(2/\sqrt{n})(s)$ =

For CIM = $(1.78/\sqrt{n})(s)$ =

n	$f_{k,n}$
2	0.3
3	0.35
4	0.37
5	0.38
6	0.39
7	0.40
8	0.40
9	0.40
10	0.41
11	0.41
12	0.41
13	0.41
14	0.41
15	0.42

Proceed to phase one, cycle two. With cycle two, there is enough data to calculate effects, standard deviation, and error limits. (See Chart 3 on page 197.) The observations from cycle one become the previous cycle sum and previous cycle average. New observations are entered, and the differences between the previous averages and the new observations are computed. It is important to be careful with the signs for the differences. The experimenter then computes the new sums by adding the new observations to the previous sums. Finally, the new averages are determined by dividing the new sums by the cycle number (n) for that phase (in this case, n = 2).

Next, the various effects are computed. Effects are calculated by plugging the new averages into the formulas on the EVOP worksheet. Then the standard deviation is calculated. In cycle two, there is not yet a previous-sum standard deviation or a previous-average standard deviation, so the experimenter proceeds directly to the new standard deviation calculation. To calculate the "new s," multiply the absolute range of differences from line "iv" (in this case, r = 2 – [–2] = 4) by the factor $f_{k,n}$ where n is the cycle number. The factor for "n = 2" is 0.3. Thus, the "new s," "new sum s," and "new average s" are 1.2.

The final calculations are for error limits. By using the factor $f_{k,n}$ we have automatically set our error limits at the 95% confidence level. For calculating error limits, plug the "new average s" into the worksheet formulas.

Analyze the results of the cycle. This consists of a comparison between the effects and the error limits. The "a," "b," and "ab" effects are all greater than their corresponding error limit of 1.7. From this, it can be concluded that the effects are statistically significant. In other words, the effects are too large to have been caused by random error. The absolute value of the "CIM" effect is less than its error limits and is thus ignored.

Since the effect that causes the greatest improvement in run length is the "ab" effect, the experimenter must then determine what is to be gained by making this the new center point for the next phase. The center point for phase one yielded an average of 89 (890,000 impressions per plate) whereas the "ab" point yielded an average of 94.5 (945,000 impressions per plate). The difference of these two is 94.5 minus 89, or 5.5. This 5.5 is compared to the error limits for "new averages," which is only 1.7. Since the net gain of 5.5 (55,000

Chart 3

Two-Variable EVOP Worksheet

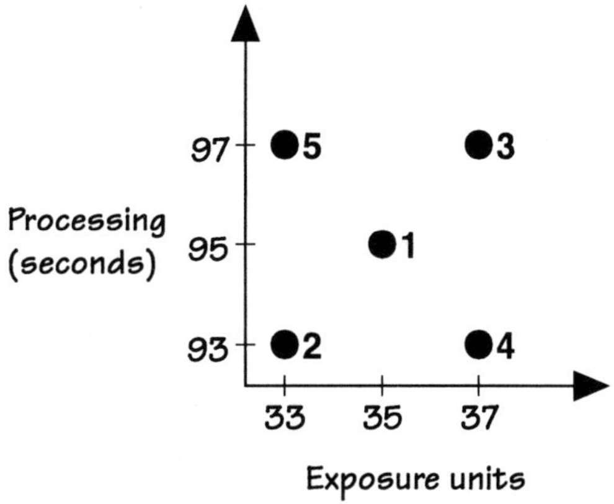

Project Example

Phase 1

Cycle n = 2

Response run length

Date 4-12-1995

Calculation of Averages

Operating Conditions	1	2	3	4	5
(i) Previous Cycle Sum	90	85	95	85	83
(ii) Previous Cycle Average	90	85	95	85	83
(iii) New Observations	88	87	94	85	84
(iv) Difference (ii – iii)	2	-2	1	0	-1
(v) New Sums	178	172	189	170	167
(vi) New Averages: y_1	89	86	94.5	85	83.5

Calculation of Effects

"A" effect = $1/2(y_3 + y_4 - y_2 - y_5)$ = 5

"B" effect = $1/2(y_3 + y_5 - y_2 - y_4)$ = 3.5

"AB" effect = $1/2(y_2 + y_3 - y_4 - y_5)$ = 6

CIM effect = $1/5(y_2 + y_3 + y_4 + y_5 - 4y_1)$ = -1.4

Calculation of Standard Deviation

Previous $\sum s$ = n/a

Previous average s = n/a

New s = (range)($f_{k,n}$) = (4)(0.3) = 1.2

New $\sum s$ = 1.2

New average s = 1.2

Calculation of Error Limits

New average = $(2/\sqrt{n})(s)$ = 1.7

New effects = $(2/\sqrt{n})(s)$ = 1.7

For CIM = $(1.78/\sqrt{n})(s)$ = 1.5

n	$f_{k,n}$
2	0.3
3	0.35
4	0.37
5	0.38
6	0.39
7	0.40
8	0.40
9	0.40
10	0.41
11	0.41
12	0.41
13	0.41
14	0.41
15	0.42

impressions) is greater then the error limits of 1.7 (17,000 impressions), the experimenter makes the decision to move on to phase two.

Proceed to phase two of the EVOP. The ranges for phase two are still plus-or-minus two, but the center point changes to the "ab" settings from phase one.

Settings for phase two: exposure = 37±2 units
processing = 97±2 seconds

The experimental procedures for phase two are identical to those of phase one.

1. Run cycle one. (See Chart 4 on page 199.)
2. Run cycle two. (See Chart 5 on page 200.)
3. Calculate averages, effects, standard deviation, and error limits.
4. Compare effects to error limits.

After two cycles, the experimenter can calculate effects and limits and analyze the results. The "a," "b," and "ab" effects after cycle two are all greater than their limits. However, this time the "CIM" effect is also greater, in absolute value, than its error limit. Recall that if the CIM is a negative value, we have reached a maximum point. Thus, at this point, we have reached an optimal level for these two input variables.

This example has shown how the EVOP moves from sub-optimal to optimal conditions. It has also shown how to run the EVOP and analyze the results. EVOP can be an extremely valuable tool for the printing industry, since there are many processes that are not yet running at optimal conditions, but are already producing a "satisfactory" product.

The following illustration shows the progression of the evolutionary operation in our example.

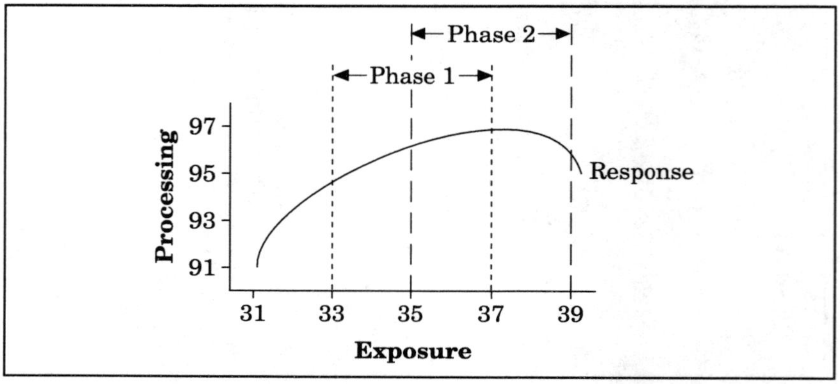

Chart 4

Two-Variable EVOP Worksheet

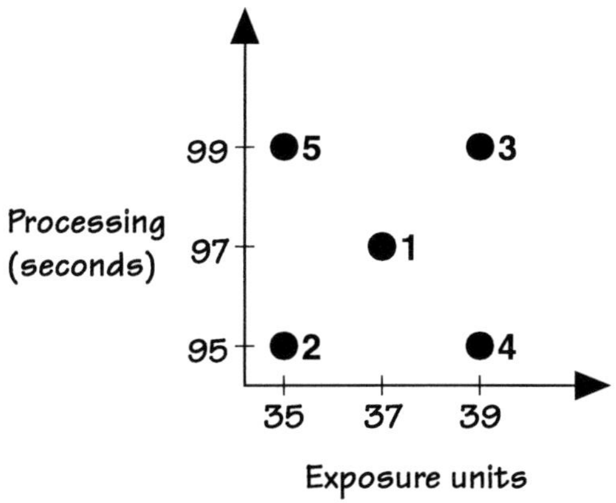

Project: *Example*

Phase: *2*

Cycle n = *1*

Response: *run length*

Date: *4-17-1995*

Calculation of Averages					
Operating Conditions	1	2	3	4	5
(i) Previous Cycle Sum					
(ii) Previous Cycle Average					
(iii) New Observations	94	88	98	89	88
(iv) Difference (ii – iii)					
(v) New Sums	94	88	98	89	88
(vi) New Averages: y_1	94	88	98	89	88

Calculation of Effects

"A" effect = $1/2(y_3 + y_4 - y_2 - y_5) =$

"B" effect = $1/2(y_3 + y_5 - y_2 - y_4) =$

"AB" effect = $1/2(y_2 + y_3 - y_4 - y_5) =$

CIM effect = $1/5(y_2 + y_3 + y_4 + y_5 - 4y_1) =$

Calculation of Standard Deviation

Previous $\sum s =$

Previous average s =

New s = (range)($f_{k,n}$) =

New $\sum s =$

New average s =

Calculation of Error Limits

New average = $(2/\sqrt{n})(s) =$

New effects = $(2/\sqrt{n})(s) =$

For CIM = $(1.78/\sqrt{n})(s) =$

n	$f_{k,n}$
2	0.3
3	0.35
4	0.37
5	0.38
6	0.39
7	0.40
8	0.40
9	0.40
10	0.41
11	0.41
12	0.41
13	0.41
14	0.41
15	0.42

Chart 5

Two-Variable EVOP Worksheet

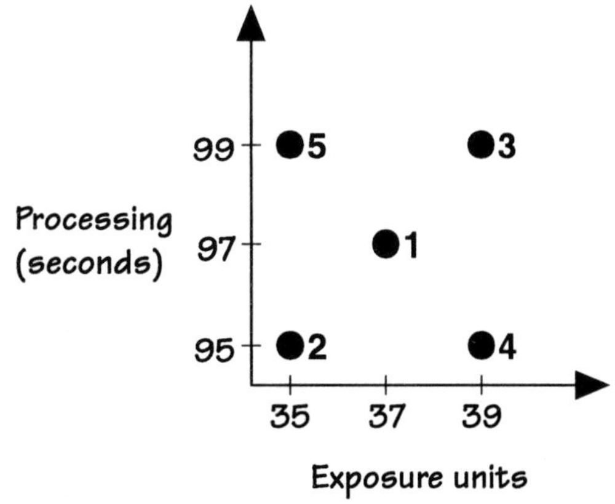

Project: Example

Phase: 2

Cycle n = 2

Response: run length

Date: 4-24-1995

Calculation of Averages

Operating Conditions	1	2	3	4	5
(i) Previous Cycle Sum	94	88	98	89	88
(ii) Previous Cycle Average	94	88	98	89	88
(iii) New Observations	97	87	96	89	84
(iv) Difference (ii – iii)	-3	1	2	0	4
(v) New Sums	191	175	194	178	172
(vi) New Averages: y_1	95.5	87.5	97	89	86

Calculation of Standard Deviation

Previous $\sum s$ = n/a

Previous average s = n/a

New s = (range)($f_{k,n}$) = (7)(0.3) = 2.1

New $\sum s$ = 2.1

New average s = 2.1

Calculation of Effects

"A" effect = $1/2(y_3 + y_4 - y_2 - y_5)$ = 6.25

"B" effect = $1/2(y_3 + y_5 - y_2 - y_4)$ = 3.25

"AB" effect = $1/2(y_2 + y_3 - y_4 - y_5)$ = 4.75

CIM effect = $1/5(y_2 + y_3 + y_4 + y_5 - 4y_1)$ = -4.5

Calculation of Error Limits

New average = $(2/\sqrt{n})(s)$ = 3.0

New effects = $(2/\sqrt{n})(s)$ = 3.0

For CIM = $(1.78/\sqrt{n})(s)$ = 2.64

n	$f_{k,n}$
2	0.3
3	0.35
4	0.37
5	0.38
6	0.39
7	0.40
8	0.40
9	0.40
10	0.41
11	0.41
12	0.41
13	0.41
14	0.41
15	0.42

Summary

As with statistical process control, design of experiments (DOE) is a means and not an end. Each should be used to complement and enhance the quality improvement process. Factorial designs and evolutionary operation are useful experimental designs that can be used by printers for optimizing the mix of key process inputs. As in other industries, printers will find it increasingly necessary to adopt these and other statistical techniques in order to effectively compete in years to come. Failure to do so could spell the difference between prosperity and bankruptcy.

9 The Team Approach

Historical Development

A major reason for the implementation of teams is the desire of printing organizations to remain competitive by increasing productivity and quality while reducing the cost of printing. Historically, however, our culture has not been built around team efforts. Most of what we have accomplished in the United States has been done because of strong individual efforts, knowledgeable individuals, trained craftspeople, and individuals looking for the quick fix to success. An understanding of these concepts will assist a printing company in successfully implementing a team culture.

Individual expertise. The history of the U.S. printing industry is rooted in individual efforts. For example many printing companies have sales representatives, estimators, customer service representatives, color separators, camera operators, electronic prepress personnel, image assemblers, press operators, bindery operators, and delivery people—all individuals performing their own separate operations. This suggests an attitude that "I alone" am responsible for the job. Printing history tells us that all of these individuals have been nurtured to be experts and isolated from the needs of the next person in the operation. Rarely has there been concern for the internal customer.

The craftsperson. In the days of the not-too-distant past, the printing industry was assured of quality printing by placing great emphasis on apprentice printer training. Printing craft unions took the responsibility of training apprentices in the necessary skills to become journey persons. This required a great deal of time and money, but the outcome was a knowledgeable labor pool. Technology changed very slowly, and so management and unions could invest in a gradual process of training. With the rapid changes in technology, we have moved away from in-depth craft skills. The mind-set of taking four to six years to perfect repetitive tasks has been replaced with people who are put on the job and have to perform almost immediately. Generally, this is done with little organized training, and it is hoped that the trainee will learn quickly without making mistakes. With the rapid technological changes that the printing industry is now experiencing, it is extremely difficult to develop craft skills that can be used without major modifications in our ever-changing printing environment. It is necessary to develop courses of study that incorporate basic understanding and hands-on skills in a environment that can do this very quickly. This lack of in-depth individual knowledge needs to be replaced by a team effort where a few individuals can share their individual knowledge to accomplish a quality printing job.

The all-around printer. Historically, the small print shop was operated by an individual (or individuals) who was capable of completing all the tasks required to sell a quality printed product. This person understood printing specifications and could sell the job, design the job, estimate the cost, do the customer service work, set and paste up the type, do the camera work, strip, print, bind, and deliver the completed job. This still exists in some quick printing operations of the printing industry and in small local print shops. Because of rapid technological changes and the time required to learn the entire printing process, it is no longer practical or possible for one person to have all the knowledge required to complete a quality print job. Because of the on-demand market and "time is money" philosophy, we are driven by tighter and tighter deadlines while maintaining a quality product at a competitive price. This requires specialization and expertise at every level of the printing operation. The "one person, know everything, do everything" printer is part of history. Now it requires a team effort to get the job done. Team efforts foster the concept that each individual will learn more about the delegated team assignment through a joint sharing of efforts than if each individual did the entire task. Teams allow people to view the entire operation without having to be experts at every aspect. The end result of the team approach empowers everyone in the printing company to work together to improve the entire process. When developing teams, it is therefore important to recognize individual efforts within the context of how they have assisted in team goals.

Quick fix. Many of our ancestors and many new immigrants came to the United States with the belief that the streets were paved with gold ready for the taking. The perception that the United States was a country that could provide instant wealth and prosperity without much work was fostered: one minute poor and the next rich. The promotion of the lottery or betting at race tracks and gambling casinos are a form of quick fix.

In the industrial world, this same mentality pervades. We see problems as one of a kind to be dealt with through the fire fighting technique: fix the individual problem, get the job out as best we can, and go on to making the same or similar mistakes on the next job. It is a never-ending cycle of quick fixes. The people who can do this the best are reverently called "troubleshooters": treat each fire as an isolated issue, deal with it, and move on to the next fire. If there are pinholes in the negatives, we opaque them out without analyzing why we are getting pinholes so that we can correct the problem and reduce opaquing time. In the short run, this technique has served to get us through the day. However, this is an inefficient and costly method of doing business. Most fires have underlying causes, which when fixed will permanently prohibit similar fires from happening. Companies need to document their fire fights, analyze the underlying causes of the problems, and schedule them for permanent extinction.

When developing teams, it is necessary that people review their progress on a regular short-term basis so that they do not feel that they are wasting their time and lose interest in the project. People must perceive that progress is being made; that is what intrigues us with the quick fix.

Considerations for Avoiding Teams

The average printing job requires the skills of a great many people to achieve its success. Print quality is a matter of understanding customer requirements; developing specifications that will meet these requirements; having materials, equipment, and methods that can perform to these specifications; and finally putting it all together in an environment that is conducive to meeting these customer specifications on a consistent basis.

It would seem obvious that a team effort is required to achieve a quality print job. So why have we not developed team efforts to achieve this goal? Some of the reasons lie in the history of the loss in craft, thinking that we will be giving up autonomy and our general culture of doing it ourselves. However, some of the resistance to team efforts is the perception that the financial "up-front cost" of team development is too prohibitive.

Up-front cost of team development. There is no doubt that there is an "up-front cost" of using teams. Teams take time to work together, to define a problem, to analyze the causes of a problem, to wade through the possible solutions to a problem, and to finally implement the solution. Teams are not a quick fix and therefore require time to resolve problems, which, in essence, is money. Many printing companies have perceived that using a team approach to resolving problems would be too expensive to the short-term profits of their company and therefore did not implement the team concept in their companies. When they looked at the initial cost of its development vs. the long-term gains, they have decided that the effort would not be positive on the immediate profit statement.

There is an up-front cost to team development. It is essential to calculate the cost of team development and estimate when the company will get a payback for this investment. Not doing a cost analysis will lead to frustration and the demise of the team concept.

When developing a budget for teams, one should look at the time required to train people for team activities, team meeting time, awards and other rewards for success of a project, consultants' time, management time, and a contingency fund for miscellaneous expenses. This figure then should be allocated as a fixed cost of doing business and allocated to the budget hour rate for all printing jobs. In this way, you will be sure that the team process will not be disbanded because of the money involved. The expense of teams of employees working together needs to become a part of the cost of doing business.

Team Costs

- Team training
- Team meeting time
- Rewards
- Management time
- Consultants' time (coach, facilitator)
- Miscellaneous

Frustration. Working through people can create an atmosphere of frustration. Based upon our history, we are not used to working and thinking as a unit. We are not used to being questioned. We are used to doing things our way. We are used to dealing with fire fighting as an every day occurrence and not thinking in terms of resolving problems so that they will not occur again. We are not used to convincing people of our point of view.

To work within a team requires patience from each individual. We must listen with an open mind to the facts and perceptions of each person. We must try to develop perspective from those around us in the team. We must learn to be patient with varying personalities. This requires a great deal of patience and can lead to frustration if each team member is not prepared for it.

Benefits of Quality Improvement Teams

Why use teams? If teams are going to "fly in the face" of our history and culture, cost up-front money, and lead to individual frustration, why would we use a team approach for quality improvement of a printing organization?

The future of printing will demand more complex expertise on the part of individuals. We will be required to rely more on each other to resolve the complex issues involved in manufacturing a printed product in a quality, cost-effective, and on-time service-oriented manner. The culture of the craftsperson will be best served if we see ourselves as part of a unit bringing our own personal expertise to the team in order to achieve our individual goals. To produce the entire printing job, individuals must be interconnected. The technology, customer needs and expectations, costs involved, and service required to fulfill our printing needs is much too complex for any one individual. We must look at printing as a system, not as a group of disjointed crafts and knowledge bases. We are each others suppliers and customers. The more we can communicate our specifications and understand and meet the needs of our customers, the greater the opportunity for successful printing jobs.

Some of the benefits to quality teams include buy-in, being part of the whole, developing a participative management philosophy, reducing the fear of change, the development of continuous improvement, understanding the structure of the printing process, developing a unity of purpose, and developing a company culture.

Buy-in. The statement that "we do not do it that way" might mean "it was not invented here" (NIH). It means the implementer had little or no input to what they are being asked to do or change. Generally, this leads to unfavorable attitudes, and most of us will find fault with the change, negate the change, and even sabotage the change. If, however, the procedure is developed by a team, all involved will be more likely to implement the change. If one is part of the study and understands why a change is necessary, the chances for long-term continuous improvement are increased. The buy-in issue is an up-front cost of team participation. People need to be involved at the beginning of any project and have input because the lack of buy-in will cost more in the long run. An idea, no matter how great, remains only an idea if not implemented.

Participative management and coaching. Getting work done through people is one of the basic objectives of management. The modern manager must be more coach-oriented. Coaching methods include teaching basic methods of problem solving and facilitating employee needs. The modern manager must make sure that the right materials and equipment are available to do the job. The people doing the actual work need to develop the methods and assist management in determining the specifications for the materials and equipment. This allows for unity of purpose where management and the operative staff are part of the same team because they share in the total effort of manufacturing a printed job.

Continuous improvement. One of the major concepts surrounding the modern quality movement is the development of an organization that is continually improving its process. This requires that people in the organization understand why they are operating the way they are, as well as how to get a specific job through the process. In the past, we have been mostly concerned with the job presently at hand and not concerned with studying and understanding why we do the job the way we do. This frame of reference precludes process improvement.

Process improvement requires that employees be aware of job flow, causes and effects of print manufacturing, items that are most likely to cause the greatest problems, and whether or not the process is in control and being maintained. The understanding and control of these critical criteria requires a team effort. Only through a team effort can we be assured that continuous process improvement of the printing operation is sustained.

Developing a Quality Team Effort

The types of quality teams. The ***executive quality council*** (quality steering committee), ***quality circles,*** and ***cross-functional teams*** (quality action or project teams) are the three major types of teams that are most commonly used by companies that are seriously involved in employee team implementation.

The executive quality council or quality steering committee. The council is a permanent team made up of representatives of top management and middle management and can also include first-line supervisors and representatives of the manufacturing work force. Middle management, first-line supervisors, and manufacturing personnel can be rotated on a periodic basis. If the organization is large enough to support quality assurance, financial, and human resource managers, these staff positions should also be part of the council.

The chairperson of the council is generally the organizational quality guru. A guru is someone who takes the responsibility to not only support quality efforts but also champion quality causes wherever and whenever possible. In small printing companies, it may be the president. In larger companies, it can be the quality assurance manager. The person who takes this assignment needs to be committed to total company quality and give this job top priority status in his or her daily work commitment. The quality assignment must be a significant part of this person's job description.

The council is a permanent team that meets on a regular, scheduled basis. The objective of this group is to formulate quality objectives of the organization. The council needs to focus on short-range as well as long-range planning. Long-range plans are those that are two to five years in duration and includes customer relations, competitive positioning, internal communications, and other holistic issues of the company.

The council must stay away from specific solutions and details of a project. Details of a specific quality issue are best resolved by experts and people closest to the quality issue involved. Councils must not tinker with how to meet the quality objective, but rather with developing a framework from which qualified employees can work. The council should focus on the process and allow specific teams to spend their time on content.

The council is responsible for reviewing and prioritizing all quality objectives, deciding on which projects to focus on and assisting in formulating a clear statement of objectives, choosing team leaders, defining team costs, and agreeing to the members of the quality action team. The council is also responsible for accepting or rejecting quality action team solutions to a given problem. If solutions are accepted by the council, they are also responsible to make sure that the solutions are implemented.

The council should have between three and ten members. Remember when you do a cost analysis and budget for quality, you need to consider the number of people and the amount of time required to do the job, both individually and as a group.

Keep in mind that a good deal of the work will be done by individuals and smaller groups. Council members need to have delegated assignments. Meetings of the entire group should be for report reviews. Individuals assigned to the council can consider that this task will require up to 10% of their time. This can be as high as four hours per week. For budgeting purposes, council team members will need, on the average, two hours per week to do their jobs satisfactorily.

The following assumptions are for estimating purposes only and will vary depending upon company objectives and internal costs. One can assume that the company will need to budget an average of five people for the council at $30 per hour for two hours per week, or a total of $300 per week. As part of the yearly budget, the council will then be financially budgeted for $15,000 for time. Add to this any clerical support necessary and other miscellaneous items of working with other employees, suppliers, faxes, and phone calls and the budget for the council can easily exceed $20,000.

Quality Council Costs

- Total number of members
- Time commitment of each member
- Secretarial time
- Hourly cost for each member
- Miscellaneous

Quality circles. Historically, quality circles were teams that were set up within a specific environment such as the pressroom or the sales department. The primary mission of the quality circle was to improve human relations of a department with a secondary focus on the improvement of quality. The scope of the projects was generally related to a particular department. The quality circle team could focus upon any useful project, not necessarily those determined by Pareto analysis, which emphasizes focusing upon the vital few problems. They were generally ongoing and permanent in structure. The quality circle was focused on departmental issues of quality. They looked for problem areas and presented these problems to the council for their approval and priority. The quality circle was not generally given any specific direction or problem to solve. It was the responsibility of the quality circle to find problems. The quality circle did not include internal or external suppliers and customers.

The leader of the quality circle traditionally was not the supervisor, but a small-group informal leader or outside facilitator. Whoever was chosen was trained in the skills required to facilitate a quality circle. However, quality circles have evolved to include department supervisors as quality circle leaders and the use of Pareto analysis as a tool to determine quality projects.

Team members could be rotated and be voluntary. However, in recent years there is a tendency to appoint members rather than have them volunteer for the assignment. To volunteer means that quality is not a company requirement, whereas to be assigned to the quality circle is to verify that quality is part of everyone's job. Companies that assign employees to quality circle teams must be aware of the problems associated with nonvoluntary assignments. People generally resent these assignments and do not work at top performance. Members of quality circle teams must involve themselves intellectually in order for these teams to be worth the time and effort. Assigning employees to quality circle teams can work in printing organizations that have been involved with employee empowerment and team efforts. Employees must observe the benefits derived from quality circles before being assigned to be a team member.

Activities of quality circle teams also need to be budgeted. Quality circle teams generally meet once a week for one hour. It is suggested that a printing company start with one team. Choose the department that is most open to team work and sharing of ideas. The department should be open to change and not feel threatened by it. Changes can be in work methods, technology, and materials. The eventual goal is to have each department have a quality circle team as part of the commitment to total quality management. This would mean that sales, office staff, prepress, press, bindery, and shipping would each have a quality circle.

It is advisable to have at least three members per group (no more than seven members per group). If you have six three-member groups and each group meets one hour per week, you will need to budget eighteen hours for this endeavor. If your average salary with fringe benefits costs $25 per hour, you must budget $540 per week or $23,400 per year.

Costs of Quality Circles

- Number of groups
- Number of members per group
- Hours per week
- Secretarial assistance
- Miscellaneous

Lessons learned from the initial development of quality circles include having the supervisor as a participant in the activities of the quality circle and the use of Pareto analysis to assist in deciding on the immediacy of the chosen project. When the supervisor was not involved, the quality circle activities became suspect and implementation was limited because the supervisor was not part of the decision-making process. The use of Pareto analysis to assist in deciding projects helps eliminate personal opinions and brings quantification and scientific justification to quality projects.

Cross-functional teams (quality action teams or project teams). The cross-functional team is also known as a quality action team or project team. For the purpose of this discussion, we will use these names interchangeably. These teams are designed to solve specific problems. The life span of the team is the length of time that it takes to resolve a particular quality issue. Team members can be used to implement accepted solutions and can go on to be part of other teams.

Members of quality action teams are chosen for their expertise in a specific area and for their job-related concerns for quality issues. If the problem is one of communication between the sales department and the manufacturing unit, we may find that the team is made up of personnel from sales, estimating, customer service, prepress, press, bindery, and shipping. The objective is to develop a team made up of suppliers and customers who are concerned with the quality issue and its resolution.

People must want to participate on the team. However, it is suggested that volunteers should be used only after potential team members have been asked to participate. Team members must have something to contribute besides enthusiasm. Each member brings expertise and perspective to the subject. Quality is not a voluntary issue to a printing company; it is part of everyone's job and therefore serving on a quality action team must be considered part of the normal workday. Quality action team members are chosen to volunteer; however, not everyone will want to participate. It is best to allow people the opportunity not to serve on quality action teams. If people volunteer and are not chosen, it is important to explain the reason why to them and to harness their enthusiasm on another team. You do not want to turn off an enthusiastic employee to team participation.

Getting Started

A checklist for continuous process improvement. The following is a checklist of procedures that are suggested for getting started in a printing company's quest for continual process improvement.

1. Chief executive officer (CEO) or his/her designee assigns quality champion (quality guru/facilitator) role.
2. CEO and quality champion organize the executive quality council.
3. Quality council develops or analyzes mission statement for the company. This may include a vision statement.
4. The council develops a total quality management (TQM) budget.
5. Quality council solicits quality project information.
6. Quality council identifies and prioritizes quality-related projects.
7. Quality council chooses team leader.
8. Quality council and team leader write a team mission statement, identify quality action team members, and develop a project budget and the time frame.
9. Team leader is trained in quality action team leadership techniques and statistical quality control tools.
10. Team leader organizes the quality action team.

11. Team leader works with team to clarify charter, budget, and time frame.
12. Team leader trains team*.
13. Team sets meeting times.
14. Team leader, with the agreement of the team, delegates specific responsibilities to team members.
15. Team members report back to team on results of their assignments.
16. Team leader keeps executive quality council apprised of the progress of the quality action team.
17. Team makes presentation to the council at the conclusion of the study.
18. Council approves or disapproves the study.
19. If the council approves the study, the council and team (possibly not all the members) implement the plan.
20. The quality team is disbanded after a "process well done" celebration.
21. Council reviews project implementation on an ongoing basis for a two-year period. It is suggested that this be done after thirty days, then six months, and one and two years.
22. Go on to next project(s) and continue the process of quality improvement through the use of quality action teams.

The quality champion (quality guru/facilitator). The quality champion (guru/facilitator) must be chosen carefully. The person chosen for this position must have the following qualifications:

1. A persuasive management style. This person will need to work with people to convince them that quality is the number one priority for everyone in the printing company. She/he also must understand politics, i.e., that which is possible to achieve in a given situation.
2. This responsibility requires time. The person assigned to this vital task must be given time to achieve results. This time frame needs to be discussed before any other action takes place.
3. A person who has a broad knowledge of how the company operates. This should include an understanding of

---

*It can be cost-effective to choose members of the first action team for their leadership ability. Then the members of the initial action team can be used as team leaders for other action teams.

customer needs and expectations, marketing and sales, as well as manufacturing.
4. This person must enjoy learning and sharing information.
5. A person who enjoys teaching.

The cost of the quality champion will vary with the size of the company. In smaller companies with up to twenty-five employees, this job will require a minimum of eight hours per week and can be combined with other responsibilities. For companies with fifty employees, this job will require a minimum of twenty hours per week. For companies of seventy-five and more, this is a full-time job.

Organizing the executive quality council. The executive quality council (EQC) should be made up of a cross section of individuals in the organization who can bring expertise from each of the areas within the printing organization. When choosing these individuals, review their responsibilities. Marketing and sales, estimating/production planning, manufacturing, and shipping/receiving should have representation. The council must include quality assurance, financial, and human resource managers, if the printing company is large enough to support these staff positions.

Developing a mission and quality statement. A company's mission statement should include how it will deal with quality issues and process improvement. It is suggested that the quality mission statement be derived from the company mission statement or revise the company mission statement to make sure it includes quality.

Mission statements and quality mission statements need to be coordinated by top management with significant input from all the company's employees. The mission statement needs to be prominently displayed. Some companies print

The Purpose of a Quality Mission Statement

- Separate the company from the competition
- Act as a driving force for continual improvement
- Give the organization a sense of unified identity
- Give the organization a sense of direction
- Be used by the entire organization as a compass or guide for continuous quality improvement.

How to Write a Mission Statement

- ***Purpose***—the company's reason for existing, marketing areas, distinctive competence, products and services, and how they are to be provided.
- ***Business aim***—the company's growth, performance, profitability, and market share.
- ***Self-concept***—how the organization wants to viewed by customers, competition, and employees.
- ***Policies***—company philosophy and leadership style that addresses the relationships of how the company is internally managed.
- ***Values***—business practices upon which the organization can be judged and should include management style, products, services, quality, and pricing structure.

wallet-size versions of their mission statement, which are distributed to employees, vendors, and customers. All the activities of the company must be congruent with the mission statement. Everyone in the organization must be aware of the mission statement and understand how their particular jobs relate to the company's mission statement.

Mission statements are dynamic and need to be reviewed on a regular, planned basis. Organizations change and evolve over time and so does their particular mission. If the mission statement is to remain a focus, everyone in the printing company must evaluate its effectiveness as a guiding principle for their individual efforts as they relate to the company's mission statement.

Generic Mission Statement
Quality Printing Company

Our company is dedicated to providing the highest-quality products and the most-reliable on-time service at fair prices by meeting our customers' requirements through constant control of our products and services while striving for error-free work and early detection of nonconformance. Through the process of teamwork, commitment, and prevention, we will make every effort to understand the requirements of our customers and be committed to doing the job right the first time.

Defining mission statement terms. Do not accept the above mission statement without raising some very pertinent questions. Most printing companies will accept the above statement and be extremely proud to post this statement as their mission. This can be a problem because it would be very hard to live up to. A mission statement must be definable by the printing company. A test of the validity of the mission statement is to have members of the company explain the critical terms within the mission statement and explain how the company goes about meeting them. The critical terms in the above generic mission statement are:

- ***Highest quality products***—How does the company go about quantifying the delivery of the highest-quality products.
- ***On-time service***—Is this 100% of the time or somewhat less?
- ***Fair price***—How is the company's price in relationship to others, and what is meant by fair?
- ***Customer requirements***—Can the company identify and quantify the customer requirements? Have customer needs been quantified and does everyone in the company understand these specifications and the procedures necessary for maintaining them?
- ***Constant control of our products***—Does the company work with suppliers to assure consistent incoming materials? Does the company use SPC tools?
- ***Error-free work***—How can this be proved? What is the company doing to assure that this is happening?
- ***Early detection of nonconformance***—When things go wrong, how long does it take to catch these problems? How many jobs have to be rerun or price adjustments made with customers?
- ***Teamwork***—Does the company have organized teams looking at process improvement procedures? Are employees fixing blame rather than resolving problems?
- ***Committed to prevention***—Is the company having less problems this year than last? Are people constantly working toward resolving problems or just putting "band-aids" on problems?
- ***Doing the job right the first time***—What communication systems are in place for this to happen? Are waste and spoilage amounts predictable and being reduced? How many reruns or adjustments are being made? Are systems in place to assure that mistakes do not take place?

Do not write a mission statement that is fraudulent. Be careful that what is said in the mission statement is true.

Everyone in the organization must understand the content of the mission statement and work toward keeping it truthful by doing their jobs in compliance to the mission statement.

Budgeting for total quality management (TQM). It is imperative that the up-front costs of quality be analyzed and calculated into a company's cost of doing business. Quality must be part of the overall strategy of doing business. It is part of a company's fixed costs and must be compared with any other financial function of doing business. The up-front cost of quality can be compared with such up-front cost areas as marketing, sales, estimating, rent, lights, and all the other fixed items of doing business in a competitive environment.

As in all fixed budget areas, the cost of quality must be considered in terms of return on investment and the cost of doing and staying in business. It must be kept in mind that not all of a company's quality commitments will bring in an up-front return on investment. It is difficult to estimate all the reasons that customers will try another printer, but we do know that poor service, late deliveries, and lack of understanding of customers' needs and expectations will drive customers to the competition. The members of the executive quality council must ask themselves, "Will we remain competitive, stay in business, decline, or grow if we budget for this quality item(s)?"

Quality is not an altruistic purpose, but a very practical means of ensuring a company's survival. To be sure, as you get caught up in the quality movement, you will find that "it feels good" to be part of an organization that has committed itself to quality printing that can be measured and verified.

The return on investment can be viewed in terms of what a printing company needs to invest in to ensure that the quality level demanded by the customer is maintained. The first chart on page 219 shows the categories that need to be defined and budgeted for before you can get started in your quality effort, while the second chart shows the items that must be considered to determine the return on investment.

Soliciting Quality-Oriented Projects

In defining quality projects, it is advisable that they be related to process improvement criteria. The quality council should issue overall guidelines for projects that are related to the following:

1. ***Customer needs and expectations***—Will the project satisfy the customer more than what is being done at

Quality Cost Investment

1. Technical and managerial training costs	$ ______
2. Preventive maintenance costs	$ ______
3. Control devices for printing (UGRA scales, color and register bars, etc.)	$ ______
4. Tools and devices for measuring consistency of quality (register gauges, densitometers, etc.)	$ ______
5. Time required to measure and record quality levels	$ ______
6. The development of internal and external supplier and customer specifications	$ ______
7. The time spent by the executive quality council and quality action teams	$ ______
Total Cost	$ ______

Return on Quality Cost Investment

1. Potential new customers because you can guarantee consistency of product	$ ______
2. Retention of old customers because no one else can do the job as well as you can	$ ______
3. Individual quality projects that lead to more profit:	
a. Fewer remakes	$ ______
b. Lower inventory investment	$ ______
c. Fewer job reruns	$ ______

the present time? Will customer expectations be better served when this project is completed?

2. ***Supplier specifications***—Do we have supplier specifications that are understood by employees and the supplier as they pertain to the end customer?
3. ***Job communications***—Will the project improve communications at any level of the organization?
4. ***Internal specifications***—Will we be able to work more closely with each other as internal suppliers and customers?
5. ***Equipment capabilities***—Will we understand our print capabilities as they relate to the jobs we now do or jobs we may wish to do in the future?
6. ***Preventive maintenance***—Can this program satisfy a preventive maintenance plan that will enable us to do more consistent quality work?

7. ***Training***—If we study quality in terms of how well our employees are trained, will we become more effective as a company?
8. ***Waste and spoilage***—Will the project assist us in reducing non-value-added waste and spoilage?
9. ***Delays***—If we reduce delays, can we more effectively service our customers?
10. ***Redoing jobs***—Will this study lead to a program of process improvement that will reduce and eventually eliminate redoing jobs?
11. ***Redoing portions of jobs***—Are there significant portions of jobs that we have to redo?
12. ***Customer retention***—Will the customer "see" the improvements from this project and therefore assist us in becoming a more vital supplier of print?
13. ***Delivery dates, both internal and external***—Just-in-time delivery dates are very cost-effective projects and allow us to better service customers.

Projects can be solicited through employee suggestion systems, which include a standardized form with information on how to apply, who will decide on the merit of the project, the type of projects being solicited, how long it will take to be informed of the disposition of the suggestion, what the award for winning is, and written statements from employees. Reviews of customer complaints, delays in deliveries, high waste and spoilage, and other perceived quality deficiencies within the organization may be considered. These questionnaires can be placed in a suggestion box. A sample suggestion questionnaire is shown on page 221.

Issues to Avoid

When requesting suggestions for process improvement, it is generally agreed that certain issues are very sensitive to employee/employer relations and should be avoided. As companies grow in the total quality management process and employees become comfortable with being empowered, the organization can consider addressing these issues through employee participation. It is important that certain issues related to human resource management not be included as part of quality action team projects. Some of these are salary, performance appraisal, and disciplinary issues.

Salary issues. Employees can form teams to decide on the skills necessary to perform certain jobs. It is also important

Casey Press Suggestion Questionnaire

We need everyone to assist in our quality mission. Do you have ideas as to how we can improve? It does not matter whether they are large or small. Please fill out this questionnaire and put it in the suggestion box.

Rules:

- The quality council will review the suggestion within two weeks.
- If your idea is accepted, you will receive a $25 gift certificate, which can be used at your discretion. We will also publish your name in the next company newsletter.
- Twice a year the quality council will choose the "Idea of the Half-Year." If your idea is selected, you will win four dinner tickets to the restaurant of your choice for a dollar amount of up to $125.
- No salaried or supervisory personnel may participate.

The quality council is most interested in pursuing suggestions that satisfy the following criteria:

- Benefit the customer
- Benefit the employees
- Low cost to accomplish
- Reduce the cost of printing
- Something that can be accomplished in less than a year

Category of the suggestion:

- Quality and customer satisfaction (internal/external)
- Safety/housekeeping/environment
- Materials/waste and spoilage/cost savings

All suggestions will become the property of our company. Good luck!

Name ______________________________ Date ____________

The suggestion ______________________________________

____

Who will be affected positively by the idea and in what way? ____________

____

Will anyone be affected negatively by the idea and in what way? __________

____

to involve employees in the writing of job descriptions. However, salary issues are very complex and personal and require a great deal of knowledge of the marketplace and company cost centers and strategies. There have been legal rulings against employee teams developing salary structures. It is important to consult with a legal advisor as to what issues may be prohibited from team efforts.

Performance appraisals. Some companies have worked out procedures for peer review. If the performance appraisal is for process improvement, then a program that identifies weak and strong points of an individual's job performance needs to be assessed in terms of not only what the individual can do to improve but also how the organization will assist in the process improvement area.

Disciplinary procedures. Employees can be involved in the determination of what disciplinary actions are to be taken in a general sense, but the actual carrying out of disciplinary actions is the purview of the company. Discipline must be done in a timely manner, consistent and with very specific rules. This is difficult when done by a committee.

Identify and Prioritize Quality Projects

The ideas generally come from the council's request for suggested quality projects as outlined in the guidelines for "Soliciting Quality-Oriented Projects." The council should be open to any additional suggestions.

The executive quality council reviews all projects and determines the worthiness of each one of them. The council must give a response to every person who has submitted a project. The policy must state that every project will get a response within a given amount of time. Anywhere from two to four weeks from the time of submission is acceptable. Whatever time frame is chosen, it must be adhered to; otherwise, suggestions will be considered a sham. It is most important to develop procedures that are followed religiously. This process is one of the first stages of employee involvement, and if the council does not respond in a timely and consistent manner, the whole procedure will fail. It will not only fail, but it will be difficult to get employees involved in future TQM endeavors. Be careful not to make the council's request questionnaire so rigorous that no one will take the time to suggest quality projects.

All projects should include a proof of need. The following checklist can be used to determine project priorities:

1. Is the problem specific and well defined?
2. Who will be positively and negatively affected by the project?
3. What is the estimated cost of the solution of this project?
4. What is the up-front cost of doing this project?
5. What is the return on investment if we undertake this project?
6. What is the estimated time required to do this project?
7. Do we have the internal expertise to do this project?
8. Can we get external experts to assist in resolving this problem?

When starting your company's process improvement strategy through the use of quality action teams, it is strongly suggested that you use the following guidelines in choosing initial projects. These projects should be low in investment for the company, have high customer impact (internal and external), should have low internal resistance from company employees (including all levels of supervision and management), should lead to results that will be highly visible to everyone, and should deal with a problem that is considered a nuisance and can be resolved in a short period of time.

Low investment. Choose projects that have a low initial investment. Buying a new press sounds great, but is quite expensive. However, finding out the capability of the press you now have and its relationship to what it produces is by far a lower initial investment and is the first step in determining whether or not you can fulfill your customers' specifications.

High customer impact. As your customers will be positively impacted by your quest for continuous quality improvement, letting them "see" what you are doing to be the preferred supplier is a great way to build enthusiasm for your organization by your customers. Perception is an extremely important part of doing business. A printing company that is perceived as a quality organization has the edge in getting and keeping customers. This is not only true of external customers, but of internal customers as well. How a properly filled-out job jacket assists manufacturing in saving time, reducing the chance of making mistakes, and reducing the frustrations of not knowing pertinent information is an important part of efficiency and profitability.

Least internal resistance. It is best at the beginning of your process improvement strategy to start with projects that most of the people feel are problem areas. Some operations, materials, and equipment have "friends" who will figuratively defend them to the death. The council should consider what changes will be brought about by studying and implementing all projects and be prepared to handle resistance in a positive manner.

High visibility. Choosing a project that has high visibility will give impetus to the quality endeavor of your company. It is important that quality projects not be hidden. Everyone will be watching to see if the company really means what it says about quality, and the success of initial projects and their visibility to everyone is very important.

High-nuisance problems. The more of a nuisance a problem is, the greater will be the effort to resolve it. Also, the elimination of a nuisance problem will greatly increase the "buy-in" to the quality process. The elimination, through team efforts, of problems related to methods, equipment, materials, and the general working environment will assist in getting people to understand the real value of the team process.

Cycle time. There are two types of cycle time. There is the cycle time of how long it takes the problem to reoccur, and the second is how long it will take to resolve and implement a solution. In both cases, it is best to choose initial projects with short cycle times. It is recommended that you choose projects that cycle in a day or two. In this way, a team can easily study the problem(s) and determine whether their solutions resolved the problem. This will allow everyone in the organization to grasp the importance of these team efforts.

Action Team Development and Training

The development of a plan for the implementation of teams is imperative; it is easy to fail when there is no plan on how to execute successful action teams. The following checklist will assist the quality council in the development of a plan for the printing company's action team process.

Action Team Checklist

- ☐ Assess the role of the team leader.
- ☐ Define the criteria for selecting a team leader.
- ☐ Develop a job description for a team leader.
- ☐ Choose a team leader.
- ☐ Develop a team mission statement.
- ☐ Determine project limits.
- ☐ Determine time limits.
- ☐ Define available assistance.
- ☐ Train the team leader.
- ☐ Determine criteria for quality team members.
- ☐ Train the team.
- ☐ Determine meeting times.
- ☐ Define the delegation process.
- ☐ Determine when and how to report progress and final results.
- ☐ Implementation.
- ☐ Follow up.

Assessing the role of the team leader. The team leader should be chosen for his/her proven ability to work with and lead people. The person should view leadership as a coaching and facilitating position. Choose someone who can easily be respected by the potential team. In addition, it is important for the leader to have an understanding of the subject area to be dealt with by the team. However, it is advisable not to choose an expert in the subject. Experts generally have strong opinions as to the causes and solutions to a problem that may get in the way of a team effort. If an expert is chosen to be the team leader, that person must be strongly advised in how not to influence the thinking of other team members.

As part of the team, he/she will be the facilitator, teacher, and coach. Team leaders must know how to teach the rules of team participation, statistical process control, and the general procedures of problem solving and the scientific method of problem analysis.

The role of the team leader includes the following: facilitator, coach, teacher, encourager, task setter, process controller, and content protector.

Choosing the team leader. When appointing a quality team leader, it is important to consider the following selection criteria:

- Choose a supervisor or informal leader from the work area where the quality team is to focus its energies.
- Choose a person who has enthusiasm for solving the quality issue.
- Choose someone who has the ability to listen and communicate well.
- Choose someone who has the ability to work with other people.
- Choose someone who can easily fit into the quality team leader job description or someone the quality council wishes to train.
- Choose someone who enjoys teaching and is good at it.
- Choose someone who has the time to lead the team.

The team leader is responsible for the general tasks that are required of a committee chairperson. Refer to the sample job description on the next page for the specifics of the responsibilities of a team leader.

It is important that the team leader participate from the beginning of a project so that he/she is an integral part of the project. The team leader must assist in the development of the quality action team mission statement and must understand fully the project objectives and limits, available resources, the development of the budget, the time frame for meetings, and when solutions need to be considered for implementation. The leader must participate in choosing the team members.

Quality action team mission statement. The quality action team mission statement clearly defines the project to be studied. It should state the objectives of the project in quantifiable terms. It should delineate the limits of the project to include money and time frames for beginning and finishing the project.

The authority that the team has should be addressed in the quality action team mission statement. Can they talk to external and/or internal suppliers? Can they order materials, fix machines, stop machines, review records, change methods, and do whatever it takes to complete the task at hand? Who is available for consultation, and what experts can they call upon to answer questions and assist with the project?

Sample Job Description: Team Leader

Job Title: Team Leader

Date: When Team Is Initiated

Immediate Supervisor: Quality Council

Summary Statement:

The team leader will be responsible for facilitating the __________ quality action team, keeping the team focused upon the subject, and working toward a positive outcome.

Principal Duties:

1. Train team members in the team process.
2. Train team members in the use of SPC tools.
3. Arrange time and place of meetings.
4. Organize and direct meetings.
5. Provide discipline and encourage full participation of team members.
6. Facilitate the process of brainstorming.
7. Prepare an agenda and follow it at each meeting.
8. Keep discussion focused on the project.
9. Keep meetings from degenerating into a complaint and gripe session.
10. Allow freedom to bring up new ideas.
11. Keep team members focused on positive discussion.
12. Keep all members involved in the discussions.
13. Summarize all the decisions reached in the meeting.
14. Draw up an agenda for next meeting and disseminate it in a timely manner.
15. Make interim assignments for data gathering for the next meeting.
16. Make arrangements for outside experts to attend future meetings.
17. Submit progress reports to the quality council on a regular basis. Reports include objectives, desirable outcomes, timetable of outcomes, cost estimates, savings, and evaluation criteria.
18. Follow up requests for assistance from team members.
19. Keep minutes of meetings (can delegate or assign).

The facing page shows an example of a statement of the problem as prepared by the quality council (mission statement) and includes team objectives, project limits, time associated with team efforts, and who is available to assist with technical information. This will be discussed with the chosen team leader before the team members are chosen and charged with developing a strategy for resolving this quality problem.

Team leader training. The team leader needs to be knowledgeable in the areas of small-group leadership. These leadership skills include how to develop an agenda, setting time limits, understanding the difference between process and content roles, eliciting the proper responses from team members, keeping personal "baggage" to a minimum, and generally focusing the team on effective conclusions. The team leader needs to learn how to effectively determine individual expertise and how to delegate assignments to these individuals.

Additionally, the team leader needs to know how to use and teach team brainstorming techniques (Chapter 5). It is important to determine the length of time that will be required for the team leader to become conversant and learn how to facilitate a team using brainstorming, the tools of statistical process control, and other problem solving techniques (Chapter 5).

It is important to train the team members before any quality project is undertaken. The TQM process takes time to learn and use, but once everyone in the organization has been trained and has used this process, it will not be necessary to train these individuals in these concepts again. This will reduce the time and energy that is wasted when individuals have no procedures to follow in how to work together as a team. It is imperative therefore to follow the guidelines set forth in team training and implementation so that the printing company can become a TQM organization.

Cost-Effective Team Training

- Train company guru.
- Company guru trains potential team leaders.
- Team leaders train team members.
- Team leaders always review team concepts.
- Team leaders train only concepts that will be used.

Statement of a Problem and Associated Parameters

Statement of Problem

When running two-color jobs, we have no specification for the density range required by the customer. This creates problems with understanding what is acceptable and what is not. This leads to customer dissatisfaction and even the loss of customers. We have also had to rerun jobs.

Objectives

Determine the customer requirements. Quantify these requirements. Determine if we are capable of meeting these requirements. Develop a process for measuring these requirements. Determine the costs involved with resolving this problem as compared to what it has cost us not to have and follow these specifications. (If known, give this figure; otherwise charge the team with determining this loss.)

Project Limits

We do not wish to purchase new equipment, and we would prefer to continue to work with our present suppliers.

Time

The project team will meet on a weekly basis, for one hour, beginning the second week after its formation. Meeting times are to be decided by the team members. This project is to be part of a normal workday. Overtime will be considered, but must be justified. Each team member will be allowed one additional hour per week for assigned, independent project work. Overtime will also be considered for independent work, but must be justified. The team is allocated four hours to become conversant and learn how to work as a team using brainstorming, the tools of statistical process control, and other problem-solving techniques. The project time is budgeted at sixty total hours.

Available Assistance

Team members should organize their needs in terms of who they will need as consultants, and the team leader will present these needs to the council. It is expected that the team will be able to utilize customer input, manufacturing personnel, suppliers of materials, and possible technical-support vendors. For project experiments and analysis equipment, personnel and materials will be available upon request.

The delegation process. Team leaders must use the delegation process as a means of getting all the team members involved in the process and to use everyone's energies to bring a project to conclusion in a timely manner. Some obvious advantages to delegating tasks are that the team leader does not have to do it all, a qualified person investigates a specific subject, and individuals can take pride in accomplishments and develop investigative techniques.

When the quality team leader delegates tasks, she/he needs to consider how much to delegate, how to delegate the task, and how to remain the leader of the group while practicing delegation. It is important that the team understand that the team leader is a facilitator and coach, not a dictator.

The following chart shows the eight steps that quality team leaders should use in a comprehensive delegation process.

Steps to Effective Delegation

1. Tell what needs to be accomplished (orally explain).
2. Explain why it is important.
3. Fix limits.
4. Answer questions.
5. Assist with helpful information.
6. Adjust problem areas (be flexible).
7. Set timetable for reporting and follow-up.
8. Put the agreed-upon task and its parameters in writing.

Understanding the roles the team members bring to meetings. The success of a quality team meeting depends largely upon the following related items: the team leader's skill at handling the interaction of the team members, the growth of team unity, teamwork, and the team members' observance of an agreed-upon meeting agenda.

It is important that team members develop an informal life within the confines of the team relationship, such as establishing a common methodology for running a meeting, a common procedure for thinking through a problem, and an understanding of how others view certain critical subjects.

Team members have other commitments to their departments and the people they personally work with. Ideas and possible solutions to a problem will be weighed in light of these commitments. Team members have different perceptions of the quality subject at hand. Each individual comes from a somewhat different background including education, training, home environment, and other areas of their lives. These perceptions can place a bias that might be different from other members of the group.

People have personal needs that come out in group gatherings. Some people do not express themselves in a group. One may think that these individuals do not have ideas on the subject when in fact they may feel very strongly or have excellent suggestions on how to proceed with the quality issue. Some need to be the center of attention, while others have a difficult time accepting any new concept.

Teams are made up of people who come with their own history and preconceived ideas of right and wrong, which can be called "baggage." They have their individual points of view and behavior patterns that must be understood and handled if a quality team meeting is to function properly.

It is sometimes evident that people do not react to the idea, but to the person who presents it. It is sometimes difficult to separate the person from the thought. It is essential that we remain aware that a good idea be treated as such and not shut off good ideas because of the person who thought of it.

The roles that are enacted within quality team meetings can be broken into two categories: content roles and process roles.

Process roles are those roles that are involved with how the meeting is carried out. Usually, the team leader will handle most or all of the process roles, but this does not have to happen and depends upon the leader's style and personality plus the abilities of the group members. How thoughts are summarized and where the team stands need to be addressed. The task setter will emphasize what still has to be done and gets the group to move ahead. The decision announcer will announce that a decision has been reached on a subject after the group has come to agreement or a vote has been taken. The traffic cop decides who will talk now and who will talk next. The encourager makes sure that people are encouraged to give their opinions on the subject at hand. The person who tries to narrow differences between quality team members is referred to as the mediator. To ease the tension, a team generally has a person who acts as the playboy/girl. This role loosens up the group with some laughter and takes the strain out of what could become a negative situation.

The ***content roles*** of quality team meetings deal with the subject matter of the meeting. The person (or persons) who makes suggestions and adds to the general focus is the initiator. Information seeking revolves around the person who asks questions to clarify what is being said. Blocking the discussion by objecting to actions offered by team members is the role that the blocker assumes. We all look to or need the expert to offer facts as they relate to the quality issue being discussed. Lastly, the destructive critic is the person who tears other peoples' ideas apart.

Both process and content roles are not the exclusive purview of any one member of the quality team. These roles can and will change from subject to subject, meeting to meeting, and within subject and meetings. The roles will switch depending upon the subject, the team members involved, and the team leader. It is the responsibility of the team leader to orchestrate the process roles and facilitate, coach, and mediate the content roles that take place in team meetings.

Criteria for quality action team membership. The following are guidelines for choosing quality action team members:

1. ***Interest***—the quality issue should be personally important to this person; i.e., the problem generally bears directly upon her/his job.
2. ***Expertise***—the person chosen should have expertise in this area, have insights into possible solutions, and have experience with the subject matter.
3. ***Control***—when a solution is agreed upon, this person can assist in its implementation.
4. ***Accessibility***—the people chosen for the quality action team can easily get relevant data and have enough experience with the process to understand which data is most relevant.
5. ***Personality***—someone who enjoys sharing information, solving problems, and working with other people.
6. ***Problem solving skills***—people who have training in brainstorming, statistical process control methods, and other scientific procedures should be considered for team membership. When this type of person is chosen, the cost of team training is greatly reduced. The biggest problem with this type of selection is that it reduces the effect of company-wide expansion of teams. It is necessary to bring eager, new people into the team approach.

7. ***Internal supplier and customer responsibilities***—team members should have a strong interest in the outcome of problem resolution. People who rely upon or need to know the effects of the problem are best suited for the quality action team.

Team members must be aware of what it takes to perform as a member of the quality action team. One should be aware that it takes time and effort to be an effective member of the quality team. However, the benefits of being part of the quality process improvement team far outweigh the effort that one must put into being a quality team member. The following is a job description for an action team member.

Sample Job Description: Action Team Member

Job Title: Quality Team Member

Date: When Team Is Initiated

Immediate Supervisor: Team Leader

Summary Statement:

The team member will be responsible for full participation in the quality action team. This responsibility includes coming to meetings on time, being prepared with the completion of delegated tasks, and learning and using team interaction techniques and SPC tools.

Principal Duties:

1. Attend weekly meetings (or when called).
2. Identify quality-related problems.
3. Analyze the relative importance of a quality problem.
4. Assist in data collection techniques.
5. Learn and use the SPC process to evaluate quality problems.
6. Learn and use brainstorming and problem-solving techniques to develop quality problem solutions.
7. Develop and assist in implementing quality-related processes.
8. Solicit input from nonmembers.
9. Volunteer for assignments and follow the delegation process.
10. Be an active quality team cheerleader.

Team training. The team leader must be sure the team will function as a quality action team. To accomplish this, the team must be prepared to interact positively and must understand the delegation process, the rules that govern quality action team meetings, and the scientific approach to continual process improvement.

It is imperative that time be given to team training. The executive quality council must allocate financial support to this endeavor; otherwise quality action teams will not be successful. As suggested above, it is recommend that a minimum of four hours of training be given before the team is ready to tackle the quality project.

Undertraining and overtraining. Basic concepts of brainstorming, delegation, and the social aspects that will bring out the best qualities of team members need to be emphasized. Do not overtrain the team. SPC tools should be taught as needed. After a simple overview of the seven tools, only go into depth on the tools that will be needed to accomplish immediate tasks. The team leader should do more than just lecture to the team about these most important subjects. It is important to use exercises that will allow team members to receive hands-on experiences so that they will be prepared to work together and individually to bring about a positive resolution to the issue that the quality team is charged with.

Teamwork socialization. The following twelve items are examples of questions that team leaders need to deal with when working with action teams. This is part of the team-building process. When completed, team members will have a better understanding of how each member perceives these issues. These areas include how individual members feel about dealing with taking directions, meeting activities, conflict resolution, development of objectives, the need for innovation, team communication, the use of job descriptions, how to use the delegation process, the many issues around quality, performance appraisals, and team spirit. After a team has discussed these issues and come to a resolution on these subjects, it will be easier for them to understand individual differences and how they must work together to resolve personality and perception issues.

1. ***Directions***—Who should give the team direction? Should it be the team leader, upper management, team members, or coworkers?

2. ***Meetings***—Who will develop the meeting agendas? Should agendas include a time line and individual responsibilities? How will the team members conduct themselves in the meeting? How do we keep personal issues out of meetings?
3. ***Conflict***—Will we work toward a consensus? When will we take a vote? How do we avoid personal attacks? When is compromise called for?
4. ***Objectives***—How will the team determine the objectives for the agenda items? Will the whole team have input into the team objectives?
5. ***Innovation***—When a new concept is put forth, how will team members treat this idea? Will the team encourage members to be creative in offering new ways of approaching the problem?
6. ***Communication***—Will the team have to worry about what members say in the meetings? Will social and non-work-related subjects be allowed in the meetings? How will team members be kept informed of what each member is doing?
7. ***Quality action team job descriptions***—Will each member have an individual job description or will one job description be sufficient for the entire team? How will individual job responsibilities be allocated?
8. ***Delegation***—How will individuals be delegated their assignments? What will be the degree of autonomy? What type of assistance can individual team members expect in carrying out their delegated tasks?
9. ***Team performance appraisal***—How will the team be evaluated? Should each individual be evaluated? When should the evaluation be done? Should the team evaluate itself? What should be the criteria for evaluation?
10. ***Quality***—How will the group determine what a quality team is? Will the team collectively determine what quality is for the specific project? How will the team measure quality of the project?
11. ***Team spirit***—What can each member do to bring the team together as a cohesive group? How can team members avoid acting as a group of self-serving individuals? What is the team leader's responsibility in building team cohesiveness?
12. ***Commitment***—What will assist team members in becoming committed to the team process? What things can team members do to build team loyalty?

Team meeting time. The quality action team is responsible for determining meeting times. It is given a time budget as part of its mission from the council. This time allocation will include the time required to train the team, team meetings, and individual delegated responsibility time. The team must decide when, where, and for how long it will meet. It is suggested that a timetable be set up to include the above time allocations and be adhered to closely. Obviously, the team needs to be flexible as unforeseen items may occur. All participants must attend each meeting. It is imperative to develop a meeting schedule that has consensus of the team members. As a starting point, it is suggested that team training take place in a four-hour block of time. Sometimes, this can be broken into two, two-hour blocks of time. Use the first two hours to teach brainstorming, delegation, and social issues, and use the second two hours for the quantitative skills of SPC. After this point, meetings take place for one hour weekly. Individual assignments should also be scheduled for one hour per week. The above may not work for each team, but each team must have a plan-do-check-act plan.

Delegation of individual tasks. Quality action teams work more effectively when each team member is assigned responsibility for investigating a specific section of the quality project. Individual team members must realize that they are on the team to work, and that requires not only coming to meetings but also doing their delegated assignment before meetings. Meetings are for the sharing of knowledge as well as decision making. If individuals are not prepared for team meetings, it is very expensive to the printing company. This should not be tolerated.

The quality team delegation process should be done at the team meeting. The leader needs to have facilitated the team into deciding what questions need to be answered and what projects will make this happen. These issues then need to be delegated to one or more members of the team at the team meeting. By delegating the assignment at the team meeting, everyone gets a chance to assist in clarifying exactly what has to be done.

Team Member Reports and Action on Team Projects

Individual reports are generally given orally, with written documentation, to the team. It is important to have written reports. If individuals do not have access to word processors, it is incumbent upon the team leader to provide secretarial assistance. Some individuals get very little chance to write reports. This should not get in the way of the desired results. When this is the case, it is advisable to have someone capable of report writing work with the person to ease the burden and to assist in teaching report writing skills.

Written reports and oral presentations to the action team should follow the format in the table below.

Written and Oral Report Format

- State the problem
- The date the work was initiated and completed
- Team member(s)
- How much time the assignment took
- Objectives (reasons for investigation)
- How the person went about the investigation
- The results of the investigation
- Any needed follow-up

Progress reports to the executive quality council. The quality team leader must keep the council apprised of the progress of the team. The report should be given orally and in writing. This should be done on a monthly basis. The report should include the information shown in the table below:

Team Leader Update Report Format

- Project
- Progress to date
- Time and money spent on the project
- Stumbling blocks
- Adherence to deadlines for completion
- Needed assistance from the council

Final presentation to the executive quality council. The final presentation to the council should be done in an official setting; the company conference room or its equivalent is preferable. All members of the quality council should be present along with the team members and any other individuals who need to know the results of the project.

The time necessary for the presentation needs to be worked out by the team as they put together the necessary information for presentation to the council. However, at least one hour should be dedicated to the presentation and question-and-answer session.

The presentation should include appropriate handouts, graphs, and charts. The team members should use visual aids to demonstrate their points. Prepared flip charts, slides, and overheads will make the presentation more meaningful. It is also easier for the council to comprehend the work of the team. Remember that a picture can say 1,000 words if it is used appropriately. The report should include the information shown in the table below:

Quality Action Team Report Format for the Quality Council

- Date completed
- Time and money spent on the project
- Stumbling blocks
- Solution(s)
- Cost to resolve
- Time to resolve
- Follow-up suggestions
- Possible quality projects leading to continual process improvement
- Return to the company in terms of money, service, and customer satisfaction

It is important that each team member participate in the preparation of the final presentation. Not all team members have the ability or want to be part of the actual oral presentation. Some members will like to speak, some write, and some do visual aid preparation; however, all the members must take an active role in the presentation. All the quality team members need to be recognized for their area of expertise and participation in the entire project and the final report to the council.

Council approval or disapproval of the project. It is incumbent upon the council to come to an independent conclusion concerning the project. The council has the ultimate responsibility for continuous process improvement of the organization. However, it is imperative that the council has a stated policy on how it will deal with project team efforts.

The council must give a time frame in which it will respond to the team as to the project's disposition. This is normally done within thirty days of the presentation. The council must state why it accepted, put on hold, or rejected the proposal.

It is important that the council be aware that putting a project on hold or rejecting it can lead to demoralization of the project team and employee team efforts in general. However, there will be times that the council will put on hold or reject a project. The important thing is to answer the team in a timely manner with adequate explanation for the decision.

Project implementation. Nothing is more exciting than to see your hard work pay off. Team members should be allowed, wherever possible, to assist in the implementation of the project that they worked on. The project implementation should be carefully monitored to see the results. It is important that everyone be apprised of the positive results that the project has delivered. This must be done on a continual basis in order to assist in the development of a positive attitude toward the process improvement philosophy of your printing company.

Celebrations and recognition. It is important to celebrate the completion of positive results and team efforts. This means that when a quality action team finishes a project there needs to be a celebration. The council may reject or put the project on hold, but the celebration must still take place. Remember, the company is developing a philosophy that states, "We believe in total quality management, which means that we follow a systematic process of doing business." The quality action team followed the process, therefore they were successful.

The recognition should be appropriate with the process. The council must think through how they will deal with project team celebration and recognition for the implementation of projects. Some ideas include team parties, dinner or show tickets, a day off, cash bonuses, gifts, plaques, letters of recognition, and company newsletter recognition. Each company must decide upon its recognition policy and make it known to all quality action team participants.

Disbanding a project team. Quality action teams do not have a life of their own. They should be disbanded at the end of the celebration. Some team members will go on to other projects. It would be a shame to train people in quality action team participation and not have them serve on other teams.

Some team members may not want to participate in another team immediately. Project time and team participation can wear people down and "burn them out."

Project reviews. The council needs to set up a procedure to review projects that have been implemented. There is a tendency for old habits to creep back into our daily work life. Habits are hard to break, and we need to monitor new systems to be sure that what was proposed has been implemented and is still in effect. Project implementation can be very difficult because it generally requires a change in the way we do business. It is suggested that thirty-day, six-month, and yearly reviews be instituted. The review assignment can be delegated to the project team leader, team member, or a quality council member.

Team follow-up using the "Plan-Do-Check-Act" (PDCA) cycle. The council needs to analyze how each team has worked. When a project is completed, a full analysis of what went right and what went wrong needs to be explored so that the next action team can learn from these successes and mistakes. Use each team as a benchmark for process improvement for the next team. A company must expect a minimum of two to three years before action teams are totally accepted by the employees of the company. Team involvement is a change in how we do our work and people require time to understand and accept the benefits of team interaction. Companies report a two-year to five-year cycle to successful implementation of this process, so do not get discouraged. Go slowly, it takes time to change to a team-oriented culture in the printing organization when there has been a lifetime of fire fighting and a quick fix culture is in place.

Summary

Because a printing job usually requires the efforts of more than one person to complete it, it is necessary that team efforts be fostered. In this chapter, we discussed the need to view the job as a total product requiring team efforts and not just small elements done by individuals. The benefits of team efforts are process improvement procedures. The chapter also discussed the different types of teams and where they are effective if properly trained. As with the other elements of TQM, it is imperative that top management give team efforts support, including adequate training time, reasonable deadlines, and the financial support to sustain their efforts.

10 Implications of Change

What Is Change?

In the printing work environment, employees are involved with changes in methods, materials, equipment, people, and the environment itself. This may include exchanging one method for another or modifying a present procedure. If a new technology is going to replace a current one, there is a transformation from one process to another. Many printers change or add new printing processes as the market and customers dictate. The addition of flexography to traditional lithographic printing, the addition of UV coaters to existing presses, and the modification or addition of waterless printing to conventional lithographic presses are some of the changes that are taking place. The revolution of digital prepress and the emergence of digital presses are examples of equipment changes that will substitute one process for another one.

Change includes exchanging, modifying, shifting, substituting, making alterations, and transforming the present environment to not only keep abreast of competition and customer needs but also to exceed these demands. No organization or individual is exempt from these dynamic forces. The salesperson as well as the delivery person will encounter changes in their work lives. Managers must not only recognize this reality but also lead the way by understanding how

Change Equals

- Exchange
- Shift
- Substitute
- Alteration
- Transformation

employees will react to change. Management must respond with a proactive plan to make change a nonthreatening, normal, and accepted process in a printing company.

TQM and change. Printing companies are involved in ongoing changes in customer needs and expectations, technology, materials, environmental issues, and educational and training issues in order to give themselves a competitive edge. How does the company take a proactive stance so that these changes can occur without reducing profit margins? The objective is to maintain or even increase these margins with as little stress as possible to the work force. TQM is a methodical approach to running a printing company. The basic concepts include employee involvement and teamwork. These concepts enable people to view change as supportive of their jobs because they are involved in the change decisions. TQM uses the scientific process, which includes quantified information and other procedures, to assist in an orderly evolution within the company. When employees are involved in continually evaluating customer needs and expectations, studying what the competition is doing, benchmarking against best practices, and reviewing technological advances, then change is more likely to come from the employees rather than from top management. When this happens, change will be embraced as less threatening and be implemented in a more expeditious manner. When TQM is implemented properly the resistance to change is minimized.

Basic Assumptions Concerning Employees and Work Habit Changes

Change is not positive or negative. When implementing TQM into the printing environment, do not expect it to be embraced by the employees. Remember that change is not positive or negative, but is looked upon with hesitation and even fear. Do not be surprised to hear comments such as, "things will never change around here," or "when is this nonsense going to end so we can get back to work." These expressed statements are indications that your quality plan has gotten the attention of people; now the quality concepts will have to prove themselves worthwhile.

It can be assumed that people will be hesitant, if not resistant, to implementing a quality control process. The obvious reasons for this hesitancy stems from the basic self-defense mechanism that change implies a different future. When an employee does not know if the future will be positive or negative, there is a tendency for the self-defense mechanism to "kick-in," creating the hesitancy and or resistance to any change in the quality needs of the work environment.

Change and what it implies to the employee. Change itself is not resisted; it is the implication of the change that is resisted. People accept and even welcome those changes in their lives that are perceived as being beneficial to them. Negative or positive responses to statistical process control methods come from how the change is approached. One must consider what the implications to the employee might be and then approach the subject based upon these assumptions. If the situation to be studied is how to maximize customer service effectiveness, the customer service department must understand that this study is not an attempt to blame them for a poor job but to assist them in being more effective as a link between customers and the company. When customer service representatives are asked to develop check sheets to track problems that they encounter with the various elements that go into the job, from order entry to final delivery, this can be misunderstood as a method of criticizing their particular expertise rather than a procedure for quantifying problems. It is important to involve the customer service department in this study from the beginning and teach them how check sheets and Pareto diagrams can assist them in identifying problem areas and finding solutions.

Does change make the job easier? If the benefit of the change will make their jobs easier, and that can be readily recognized, the change in work procedures will then be accepted more readily. However, quality improvement processes that include statistical process control generally take some up-front effort. This effort can be perceived as difficult and unrewarding unless approached as a necessary activity that will reap many benefits to the individual as they are implemented. The customer service department can take their check sheets and Pareto diagrams and brainstorm a cause-and-effect diagram to assist them in resolving their ongoing problems. In this way the printing company will develop continuous improvement through employee involvement. It is important to emphasize that these processes will make the job easier.

Reactions to change can be stated positively and yet be negative. Some people express hesitation and skepticism in a positive fashion, but the fear of change still exists. Comments such as, "yeah, it is a great idea, but now is not the time to develop quality teams," or "let's wait until things slow down before implementing statistical process control" are both statements that can be interpreted as cautiously positive. However, they can stop progress as easily as overtly negative statements about the quality improvement plan. If a printing company is trying to review how to reduce waste and spoilage in the bindery, but the bindery is working at full capacity, it is incumbent upon the facilitators to insist upon a plan-do-check-act program from the bindery. This PDCA program would be built around full capacity and include a plan for the investigation of the causes of waste and spoilage, how to go about eliminating them, a strategy for implementing these changes, and when a review would be done of how the plan and strategies were reducing the waste and spoilage. Do not accept vague promises. Insist upon a plan that includes dates and times and hold the bindery accountable.

Perceived Threats

Record keeping can be perceived as a personal threat. Statistical process control is a quantified system and requires record keeping. If employees feel that the company has a history of placing blame rather than resolving problems, then record keeping will probably be perceived as a means of controlling individual efforts. The company must emphasize problem resolution rather than finding fault. If press operators are asked to keep run charts of their press operation, it must be explained that this is not to control their progress but to assist them in determining what periods of time are more productive and why. Employees must realize what is not recorded generally does not get fixed. When people feel that record keeping will be used to control their work environment, it can be assured that the implementation of these quality plans will be sabotaged. Employees must not only be involved in the record keeping process, but in the analysis of these records and what they mean and how they can be used to improve the work.

Peer group pressure. As companies become more aware of the need to ensure consistent quality in order to remain competitive and survive in a customer-driven marketplace, employees must change to "quality first" work procedures.

If the image assembly department uses statistical process control methods to better understand their processes and verify their quality, it is more likely that the pressroom and bindery will uses these methods as part of their quality verification effort. People have a need to keep up with their peer group. When one printing company embraces the TQM process, it is very likely that other printing companies will follow suit. If statistical process control is considered the "thing to do," employees will see this work environment change as positive and accept it more vigorously. People like to envision themselves as doing business in a modern manner, and this type of peer pressure not only encourages individuals in a department to use statistical thinking to solve problems but also encourages the entire company to do the same.

Threat to job security. People resist change in their interpersonal and work environment that they perceive as a threat to their job security. If people feel that a proposed change will take away, control, or diminish their importance in the workplace, this change will be adamantly resisted. TQM adds control to the individual employee. Estimators who have more control over the computer programs they use to develop their estimates will feel less threatened by these technologies and the possibility of job loss than those who have these processes thrust upon them. If the estimator can now produce more estimates in a given amount of time, this does not mean that fewer estimators are needed. It can mean that they can now service sales in a more appropriate time frame and that sales can ask for changes without imposing an undue burden on the estimator. An estimator's job can now be enlarged to include the development of budget hour rates, which gives the estimator an opportunity to become a more valuable employee. Be aware that changes in the work place can be perceived as a threat to job security.

Job status and its implications. Job status must be considered when implementing change. If employees perceive that they will have more status after learning and using the concepts of statistical process control, one can expect a more accepting attitude toward the change. The employee who perceives that statistical process control will bring more status because it is a scientific method is more likely to accept and embrace this procedure. If employees see statistical process control as making them more vital to the organization, they will be more likely to accept this quality process improvement methodology. If this is coupled with the peer group perceiving that this employee is more knowledgeable and should be consulted on how they can implement and use these quality concepts in their jobs, then one can expect that statistical process control will be readily and even eagerly accepted. However, if employees feel that statistical process control inhibits creativity and even diminishes their craft freedom, then acceptance and use of quantitative procedures will not happen. The printing industry historically has been developed by craftspeople who "know" by feel when the job is right. When introducing quality control methods to the traditional craft areas of prepress, press, and bindery, it is best to treat the subject as a skill that operators need to add to their present skills in order to do their work effectively.

Implications of inferior work. The introduction of quantitative procedures can imply to the employee that his/her work is being done in an inferior manner. It is critical that employees realize that this new methodology to quality improvement will improve their position in the work environment and does not imply that previous work performance was inferior. If salespeople must keep records of which customers reject work or demand reductions in the price of the job, the salesperson might feel that they are not doing their jobs correctly. However, the reason for this type of record keeping is to understand which customers may need assistance in understanding the importance of proper specifications. Sales departments must understand that the use of quantitative procedures assist the whole organization in satisfying customer needs.

Close supervision. The use of statistical process control can be perceived as a method to enable management to more closely monitor individual and department activities. Do employees feel that the message is that they need to be watched? Asking people to chart, record, measure, and analyze their work output can be extremely threatening. If run or control charts are used to understand pressroom activities, people need to know exactly how the chart will be used. People must be made comfortable in feeling that quality control tools will give them more control over their work environment. It must be made clear that it is not the company's intent to watch over the employee's work but to allow the individual more autonomy by understanding the particular printing process. Supervisors must make sure that they do not question the information in terms of pointing to a culprit, but as a means of learning what took place and assisting in the process of understanding and eliminating problem areas.

Freedom to do the job. If change is perceived as limiting the present individual freedom of how people do their jobs, then the acceptance of a quantified methodology to quality improvement will be resisted vehemently. Workers need to understand that measuring and recording will give them greater freedom in their work environment. Employees will be able to better understand where they can personally assist in making job-related changes to improve quality and output. Team efforts mitigate the concept that each individual knows the best or only right way to do certain tasks. If pressroom personnel learn to work in teams to improve makeready and run procedures, it will make their work easier. Each person needs to share his/her expertise so that everyone can improve the end product.

Displacing jobs. Will the implementation of total quality management eventually displace jobs? Can the introduction of a quality improvement process be perceived as a means of streamlining the organization so that there will be fewer jobs? If adoption of the quality improvement process is perceived in this fashion, the employees will do all in their power to sabotage the quality effort. If the organization changes the role of supervisors from controllers and inspectors to facilitators and coaches, how will this effect their job security? Do teams need supervisors? When a company flattens the organization and reduces the number of supervisory layers, what happens to these people? One must be careful to show how an increase in quality will lead to more job security and possibly a growth in the organization. Supervisory personnel will be even more important because they have the skills to train, assist in reviewing and improving methods, and look for improved supplies. If customers develop more confidence in the quality of the products and services they receive, it is probable that there will be more work in a total quality management organization.

Employee perception of company intent. Are employees at peace with the company? Do they feel that the company is working for everyone and that everyone is important to the success of the organization? Is your company culture one of "I will tell you what to do and you do it"? Does the organization lay people off at the first sign of a slow period or does it have a plan to do training and maintenance during slack periods? Does the organization communicate the financial well-being of the company so that employees can better understand how their efforts make the company successful? When employees feel that the company is thinking of their best interest, changes will be more readily accepted and not perceived of as a threat to job security. Companies that have a history of employee awareness do better at implementing a change in quality measurement. If employees feel that the company is not concerned with them as human beings and is only interested in an immediate return on investment, then change will be seen as a ploy to quick profits and not a means of company long-range improvement.

Change as a right of management. Do employees feel that change of any type is fair and a legitimate right of management? If management acts as a coach and mentor to the quality improvement process, employees will feel that it is a legitimate right of management to implement this process. However, if it is seen as a mandate to dramatically change the way workers perform their individual jobs, then beware of the repercussions from the work force.

In the printing industry, we generally work with artists, craftspeople, and professional people who take pride in their printing-related skills and assume that management will not interfere with these skills. It is assumed that management will not mandate work procedures and will allow individual skills to dictate how the work will be done.

The implementation of total quality management necessitates a team effort that directs efforts into best practices and not necessarily individual variations. Management needs to assert the right, through TQM training, to have people find the best procedures and processes that integrate the best practices of each individual.

When re-engineering is necessary, such as in a prepress department, it is important to evaluate the skills of each individual and assist them in developing the new skills necessary to perform the new work. How can the present skills of an image assembler be integrated with computers and the use of digital information? As computers and the use of digital information become more widespread in the printing industry, there will be a need to have everyone trained in the use of these new technologies. It will assist your organization if your employees have a sense that the company not only has the right to do these things but that it is in their best interest to embrace these new skills. Companies that take a proactive position in developing their work force will have an easier time instituting change because people will accept the rights of management as a means of improving their personal lives as well.

Facilitating Change

Group commitment. Group decision making commits each member of the group to carry out the agreed-upon decision. The feeling employees have that "we developed this plan" or "it is my idea" is the most powerful means at a person's disposal to readily accept and promote changes. The introduction and implementation of, for example, teams and/or statistical process control methods need to be embraced by the entire group for it to have meaning to the printing company.

Change and quality teams. Some of the negative statements that can be heard concerning quality team efforts and their relationship to quality-related issues may be because of the newness of the quality team process. If your organization has not used a team approach to problem solving, the concept of teams may be looked upon critically. In organizations that have tried various programs of quality improvement, employees can be expected to feel skeptical about the initiation of "another" program. The use of quality teams may fit into this category.

Group decision making. It is important to involve employees in how to implement a statistical process control plan. Employees must be active participants when introducing statistical process control in order to enhance their sense of control over their work environment. The "not invented here" syndrome can be detrimental to the implementation process. If people feel that they did not participate in the development of a process improvement plan, it is more likely to be rejected. People must take charge of making the quality plan work in their particular jobs in order for it to become meaningful. This is made easier when people feel they have a part in the decision-making process of how statistical process control will be implemented in their company. If it is important to understand the steps required to ship various types of printed products, then teach the shipping department how to develop a flow chart of these activities and then coach them through the process. The shipping department then has made the critical choices in deciding how they presently ship various printed products. It is an easy step to now coach this team on what they could do differently to make their jobs easier and possibly smooth out the shipping process.

Feedback systems. Members of quality teams have to be aware that not all the changes they recommend will be acted upon. However, management must develop systems for feedback and a working relationship with quality teams so that the work of these teams is meaningful.

Statements such as "no way, we do all this work and nothing is done about it" need to recognized as indictments of the quality feedback system. This type of statement from quality team members is an indication that your organization may have a problem in implementing employee ideas. These types of remarks can negate all the fine work your company has done in implementing a quality team environment.

Act upon ideas as soon as possible. There needs to be a system in place that everyone can easily understand. Explain why you may not choose to act upon a specific idea. It is essential to communicate both an action plan for quality improvement and a non-action plan. Employees must know why their hard work is being accepted or rejected and that their suggestions are not only important but vital to the quality process improvement plan.

Two-way communication. When implementing statistical process control, it is imperative to have a two-way dialogue with all employees. Two-way communication includes teaching the importance of the process, why it is being implemented, and what it will do for the company and its employees. Two-way communication is extremely important to the successful implementation of the printing quality process.

It is important to make sure that all questions concerning the plan be answered. It will be necessary to initiate the two-way communication early in the planning process and reaffirm that all employees understand what is happening and how they will be involved in the quality plan. Make sure that all questions are answered to the satisfaction of the employees and that you follow up to make sure that any additional questions that will surely arise are answered.

Involving informal leaders. Every company has formal and informal leaders. It is understood that communication comes from the formal leaders. However, informal leaders play a large role in the acceptance and rejection of company improvements. Informal leaders are those employees who generally do not hold official titles such as foreman or forewoman but are held in high esteem by their peer group. Allow the informal leaders to present the process improvement plan to the rest of the employees. Informal leaders are looked to for leadership in the acceptance and rejection of new ideas. In developing a quality control plan, it is best to recognize these employees and keep them involved. The process will work best if you identify the informal leaders, explaining to them the benefits of quantifying quality printing and training them in the use of statistical process control procedures. The use of informal leaders to assist in the training, coaching, and facilitating roles of the implementation process will reduce hesitancy and undue stress on the part of other employees.

Involve people who enjoy working with numbers. Most individuals have a hesitancy concerning the use of data collection, the organization of this data, and the use of this information for process improvement. This is in part due to the lack of confidence in their ability to work with numbers. Companies need to identify those individuals who enjoy working with numbers. People who enjoy and are good at collecting data, keeping records, and charting will resist statistical process control to a much lesser degree. These individuals find this type of work activity fun, and anything that is perceived as fun on the job will be accepted with minimum hesitancy. It also allows all employees to understand that the company identifies people with specific skills as very important to the organization.

Change perceptions/old vs. new. People perceive change in relationship to what it will do to or for them in terms of their work environment. Traditionally, quality was measured in subjective terms in the printing industry. Rarely were numerical values placed on the various elements that represent quality. Can statistical process control replace these qualitative old symbols as to how quality is to be evaluated? One must realize that subjective evaluations will be replaced with objective ones. The customer is the only one who should take a subjective view of the printing, and this subjective view needs to be quantified into specifications that can be followed by printing employees. This change may affect how employees view their method of doing their job. A person who thinks that they have the greatest "eye for color" may feel affronted when told that this is not enough to verify to the customer that the color remained constant throughout the run. Adding the use of a densitometer, colorimeter, or spectrophotometer to quantify consistency of color must be done with great care to ensure that the person with the great "eye for color" uses this device as a verification tool and does not feel threatened that his/her skill is being diminished

The old perception of personal craft skills needs to be used along with newer quantified skills. When combining the subjective skill of the eye with the quantified device of the densitometer, the printing company can truly claim that it has combined traditional craft-subjective qualities with quantified methods that assure a quality print job. Allow these individuals who have developed these skills to teach others how they can use these procedures to verify their quality work and identify problem areas. This procedure can become infectious and widespread for all employees.

Economic incentives for change. Economic incentives are a major method of rewarding employees who accept and use statistical process control methods. Employees need to realize that quality improvement will bring new customers and printing jobs to the company. More work will mean more security and salary or fringe benefit increases.

When employees do their jobs right the first time, it will mean customer satisfaction. This in turn can mean bonuses or other monetary incentives to employees. Employees of a company who have experienced layoffs and shorter hours due to lack of work may see statistical process control as a method of assisting customers in receiving more consistent quality. This in turn may bring in new customers and jobs as the word spreads concerning quality consistency of the printing company. To employees this means a regular paycheck. All of the above are financial incentives that need to be pointed out to employees when implementing a quality improvement process.

Making change continual. When implementing the statistical process control process, it should be done with a "well thought out, slow as you go" continual improvement plan. Think your way through each concept of statistical process control before working on an implementation process. Expect to learn as you go. This works well when a teamwork philosophy is used to develop and implement the plan. Employees need time to digest the concept of quantifying quality and have to become comfortable with the concept that each step will be analyzed thoroughly before the next step in the process improvement plan is initiated.

Teach and use one or two tools of statistical process control at a time. Allow employees time to learn the tools and use them. This does not mean that management will abrogate their responsibility to continually improve quality through a quantifying process.

Unfreezing attitudes. Attitudes toward new work procedures may be viewed with skepticism. These attitudes need to be unfrozen. Some people have fixed stereotyped attitudes toward change, and this may include attitudes toward statistical process control. You may hear comments such as, "statistical process control is a waste of time," "we can have quality without measuring, recording, and analyzing our work," "statistical process control must have been developed by some egghead," and "we have to get the job out and do not have time for statistical process control." These comments are common and need to be unfrozen. Employees need to view statistical process control as advantageous to their work environment. Look for opportunities to show how quantifying the printing run could have helped in recognizing a problem before it became costly. Getting employees to ask "how do we know this or that?" assists in getting people involved with quantifying their responses. Group decision making on how to implement statistical process control will assist in unfreezing these attitudes and resolve these common change-acceptance situations.

Reinforcing change. It is very important for employees to assist in establishing new group standards. Peer group members have to decide to assist in setting up statistical process control as a new, modern procedure. If this does not take place, it is important for management to keep encouraging peer group members to take charge of the implementation of statistical process control within their particular jobs. When press personnel see the value of using a densitometer, it is important to then assist them in making it a consistent work procedure. Once a check sheet has been used successfully in the prepress area, it then must be made a part of the operation. Do not allow a process to disappear once it has been done successfully. There is a tendency to let new procedures die because returning to old habits is easier.

Induction and ceremony. After an introductory session on statistical process control has been completed and understood, a diploma needs to be awarded. This gives the employee a sense of accomplishment. Further diplomas should be issued as people become more knowledgeable and implement the various phases of the total quality management process. For further peer recognition, some companies issue two diplomas, one for the employee and the other is mounted on a wall set aside for quality issues. As employees start to use these quality tools, further recognition is important. Do not train people, reward their classroom efforts, and then allow the process to die. It is essential to encourage implementation and reward the effort. These rewards can vary from a letter exclaiming "nice job" to a dinner for them and their families. These rewards reinforce quality attitudes and cement the change into everyday work procedures.

Five Tenets of Continual Process Improvement

Remember that change is a certainty. Change is certain and can not be avoided. One must always be on the lookout for better methods and continual quality improvement. We must be careful not to make changes, just to make changes, but to improve quality, reduce waste, increase efficiency, and generally improve the quality of life within the company.

Organize. Organize your thoughts and the initial strategy for the implementation of statistical process control. Understand what the initial changes will be, collect facts, and simplify wherever possible to expedite the change process.

Employee involvement. Involve employees in the change process by utilizing participative management. Encourage employees to share their ideas and suggestions to improve the quality implementation process. Develop quality teams. Train leaders in the process of quality team leadership. Allow employees time to assist in the improvement of their jobs by analyzing printing-related problems and making suggestions.

Emphasize the benefits of statistical process control. Statistical process control needs to be implemented gradually. When introducing the process improvement plan, emphasize the benefits to the employee, the company, and the customer. Discuss objectives and how they will be evaluated. Everyone needs to understand that the company's goal is to continually improve quality by making gradual changes to the process. Employees need to understand that quality improvement changes will be an ongoing, well-thought-out process at your company.

Support the process of quality improvement. Develop a financial budget that will allow for quality improvement. Get rid of the outdated notion that quality is something that happens when we have the time and money. Build time for quality into your budget hour rates just as you build in mortgage and electrical bill payments. Develop cost estimates that take into consideration the time necessary to make sure that the printed product will be produced correctly the first time and to the customer's specifications.

Summary

Change is inevitable and can include new technologies and work methods. How well employees accept change and grow with it depends to a large part on how it is dealt with by managers within the printing company. Without guidance from management, employees look upon the implications of change with fear and resistance. The company has to deal in a positive manner with employees so that they know that change can bring about benefits to them in their work environment. This chapter has dealt with some of these areas of change and how they can be used in ways that bring about improvements to the finished product, while also bringing about positive reactions by the workers to their place of employment.

11 The Supplier Certification Process

What Is the Supplier Certification Process?

Dr. W. Edwards Deming's fourth obligation in his *14 Obligations of Top Management* states, "Reduce the number of multiple source suppliers. Price has no meaning without an integral consideration for quality." This chapter amplifies upon this obligation and develops a systematic approach that can be used to reach the goal of consistent quality materials, partnerships, and a mutual commitment from all of your key suppliers.

There is no job in the printing environment that is immune from internal and external vendors and customers. We are all responsible for understanding the specifications of our customers. We must deliver quality products and services that meet these specifications. All vendors and suppliers are bound together as the end product relies on everyone doing their job as precisely as the specification calls for. The relay race is very similar to the race we run in manufacturing a printed product. Each person must do his/her part as the end product relies on all the pieces coming together precisely as prescribed in the specifications.

This chapter uses certain words interchangeably. Vendor and supplier have the same meaning. Products, supplies, materials, and services mean the same thing within the context of the supplier certification process.

Supplier certification is the ultimate goal. This requires a complete understanding of the needs, requirements, and specifications of the customer by the supplier and that the supplier has proven, over time, that it meets these conditions. Certification implies a partnership in which there is complete trust and faith that the supplier will continuously work not only toward meeting the customer's needs, but truly understand the customer's business and stay abreast of the changing needs of the customer.

The ***supplier certification process*** (SCP) is a vital part of the total quality management process. In actuality, it is one of the first items that needs to be addressed in the quest for process improvement. Suppliers that can not meet customer specifications will cause defects in the finished printed product. There are two types of suppliers, the internal and external suppliers.

Internal and External Suppliers and Customers

Internal suppliers are those departments and people who supply us with the goods and services that we need to do our jobs within the company. For the printing estimator, it is the salesperson who must deliver the appropriate information so that the job can be properly estimated. For the estimator, it may also be an accountant who supplies information as to budget hour rates and historical data on similar types of printing jobs.

The estimator must also rely on external suppliers for appropriate supplies and services in order to do the job in a quality fashion. Suppliers of paper, ink, plates, color separations, and other needed materials that go into manufacturing a printing job have to be able to supply accurate cost figures concerning their products. These cost figures must take into account supply specifications, order quantity, and date of delivery. If the estimator is using a computer-based estimating program, it is necessary for the supplier of the computer software to continually keep the estimator apprised on software updates and better ways to manipulate the computer package.

The form on page 259 entitled "Conformance to Specifications for Internal Suppliers and Customers" is filled out by both the internal supplier and the internal customer. The form remains with the job throughout the life cycle of the particular printing job. The form is in triplicate so that the supplier and customer can keep a copy for reference and the original copy is kept with the job jacket by the main office for quality auditing and improvement purposes.

Analysis of Present Supplier and Preferred Supplier Situation

As a starting point in the external supplier certification process, it is advisable to evaluate suppliers using the "External Supplier Questionnaire Form" on page 260. This form has been developed with specific questions to assist you in determining how these suppliers comply with the needs of the organization.

Conformance to Specifications for Internal Suppliers and Customers

Name of Supplier ________________ Name of Customer ____________________
Date Supplier Completed Work _____ Date Customer Used Supplier Work _____
Describe the Supply ____

To Be Completed by Supplier:	Yes	No
Are there specifications for the job?	☐	☐
Are these specifications correct?	☐	☐
Did you verify that you met specifications?	☐	☐
Did you receive your supplies and/or services with specifications?	☐	☐
If not, explain? (quantify when possible) ________________________		
If a problem occurred, how did you resolve it? ____________________		
Is this problem permanently resolved?	☐	☐
If not, what has to be done to permanently resolve it? ____________		
All out-of-specification items that have been shipped have been discussed with the internal customer?	☐	☐
Internal Supplier Signature ____________________ Date __________		
To Be Completed by Internal Customer:		
All supplies have been checked, before using, for conformance to specifications?	☐	☐
All specifications have been met?	☐	☐
Supplier has been notified that specifications have been met?	☐	☐
Items out-of-specification have been corrected?	☐	☐
All out-of-specification items have been discussed with supplier?	☐	☐
Customer Signature _________________________ Date __________		
Management Review:		
Quality Reviewer ____________________________ Date __________		
Was action taken on out-of-specification areas?	☐	☐
Was everyone involved informed of met specifications or corrective action taken?	☐	☐
If a new specification or procedure has been developed, has it been permanently documented and being used?	☐	☐

External Supplier Questionnaire Form

Supplier name ____________________ Date ________________

Supply ____

	Yes	No
1. Materials arrive in a timely manner.	☐	☐
2. Correct quantities are received.	☐	☐
3. Defective (out-of-specification) supplies are not sent.	☐	☐
4. Must we inspect these supplies carefully before we can use them in the printing process?	☐	☐
5. Our specifications are met by the supplier.	☐	☐
6. Are we forced on occasion to use out-of-specification materials?	☐	☐
7. Rework is not necessary because of poor-quality materials.	☐	☐
8. Large inventory of supplies are not needed.	☐	☐
9. Just-in-time deliveries are the norm.	☐	☐
10. Supplies are consistent from one batch to another.	☐	☐

Desired Outcomes of the Supplier Certification Process

Improved quality is one of the most important outcomes of the supplier certification process. However, one can also expect that material costs will be controlled if not actually reduced, service will improve, and the overall quality of the finished print job will be enhanced. When the quality of the materials that go into a print job are consistent, the cost of printing the job will be reduced. This in turn improves profitability. When the above criteria are met, it increases the likelihood that customers will be served satisfactorily and return for additional work. The assurances of quality supplies will also allow the printing company to feel secure in promising and delivering quality products. When a company consistently produces predictable quality printing, it inevitably is perceived as a "high-quality organization" and receives more work.

If the printer gets inconsistent materials, the best that the printer can get is inconsistent end results. The phrase "garbage in, garbage out" applies to supplies as well as computers. Poor and/or inconsistent materials supplied to the printing company can only lead to inconsistent quality of the end product, the printing job.

Vendors who improve their vendor material specifications because printers demanded certain specifications have a greater opportunity to improve their processes and the product they deliver.

Suppliers who partner with printers who use total quality control concepts are encouraged to use total quality control processes to improve their manufacturing process. They will in turn be better able to provide the printer customer with high-quality products and terrific service at competitive prices.

Another outcome of the total quality control process is the reduction of inventories. This happens as the cost of processing rejections goes down, adjustments for poor quality supplies are reduced, and the time involved in fixing and making do with poor-quality supplies disappears. The printer will no longer need to carry excess inventory for quality reasons.

Services Offered by the Supplier

Service is a part of determining the overall worth of a supplier. Service should not be left out of the equation when deciding who to choose as a supplier. Service is closely related to overall price and quality.

Some areas that you must consider when determining the service offered by your supplier are:

1. On-time deliveries
2. Assistance in monitoring inventories
3. Keeping you up-to-date with new products and industrial information
4. Offer "how-to-use" training programs
5. Offer technical literature
6. Offer facility tours
7. Well-trained technical support
8. Keep you informed on their ongoing quest for quality
9. Keep you informed on their process improvement achievements
10. Assist you with quality training

PDCA Cycle and Its Relationship to the Supplier Certification Process

The Plan-Do-Check-Act (PDCA) cycle can be used as a method of instituting a supplier certification program. The following is a format to follow when using the Plan-Do-Check-Act cycle for supplier programs. (EQC is an acronym for the executive quality council, and SCP is an acronym for the supplier certification process.)

Make a *plan*:

1. Executive quality council defines suppliers
2. Develop supplier certification process goals
3. EQC chooses the SCP team leader
4. EQC and team leader write SCP team mission statement
5. EQC and team leader choose team members
6. Team leader and team develop implementation strategy

***Do* the plan** (SCP team):

1. Review suppliers
2. Review specifications
3. Define pertinent specifications
4. Develop and implement supplier survey

***Check* the plan** (SCP team):

1. Review supplier survey
2. Review all specifications
3. Verify specifications with supplier capability information

Take *action* (SCP team):

1. Recommend suppliers to EQC
2. Recommend pertinent specifications
3. Recommend SCP meeting procedures
4. EQC put SCP procedures into effect
5. EQC review SCP and appoint SCP (new or old) team for further refinement

The Four Phases of the Supplier Certification Process

There are four phases to a complete supplier certification process. It is proposed that printers use this methodology as a process of improving relationships with suppliers so that all materials that are received meet the proper specifications for use, are on time, and contain the correct quantity. This four-phase process outlines a framework to assist in the development of a supplier certification process plan for a printing organization.

Phase I—Supplier Improvement Strategy

Management commitment. The first step is for management to develop a commitment to the supplier certification process. Many companies choose their suppliers by price alone. Some have committed to suppliers on the basis of service alone, and many printers have suppliers and do not know why they have selected them. When asked why, they respond “because we have always done it this way.”

Four Phases of a Supplier Certification Program

Phase I	• Supplier improvement strategy • Management commitment • Supplier certification process team • Cost/benefit analysis • Specification development • Supplier selection • Supplier awareness
Phase II	• Supplier survey
Phase III	• Supplier assistance • Supplier corrective action • Supplier assistance activity
Phase IV	• Supplier certification • Supplier recognition • Supplier councils

Although price and service are very important components of what a preferred supplier commits too, there are other ingredients that also need to be considered when choosing a supplier in order to guarantee a continual process improvement strategy for your company.

Some of the items that management needs to take into consideration include:

1. ***Improve the quality of incoming materials.*** If you occasionally have to "make do" with the product supplied to you or have to return it, you need to review why this is happening. Can another supplier do better? Are you reporting this information to your supplier? What is your supplier doing about it? Has the quality remained the same, increased, or decreased? A printing company needs to work towards 100% conformance to specifications of goods and services for internal and external supplier relations.
2. ***Improve internal supplier/customer communication.*** Define specifications and communication systems that will lead to a lessening of misunderstandings and assist in lowering internal friction levels and improve the overall quality level of the end product.

3. ***Improve external supplier/company communication.*** Evaluate how company vendors are chosen and define and communicate specifications.
4. ***Consolidate the supplier base.*** Evaluate why you use certain suppliers and develop a preferred supplier system to reduce source suppliers to one or two at the most for a given material or service.
5. ***Facilitate just-in-time purchasing.*** Work with suppliers to allow them adequate time to deliver materials and services as close to your actual use as possible. This will prove to be cost-effective for both the supplier and your organization.
6. ***Develop 100%-adequate specifications.*** Specifications need to be reviewed on a predetermined basis. A SCP must allow for the continual updating, evaluating, and use of appropriate specifications.
7. ***Reduce the cost of quality.*** When the specifications for services and supplies are clearly communicated and agreed upon, the printed product will cost less because time delays, corrections, and remakes will be eliminated.

Supplier certification process (SCP) team. The SCP team should be set up as a cross-functional team to develop a strategy for internal and external supplier/customer process improvement relations. The team should include the primary purchaser (purchasing agent) and those people that are intimately involved with the product or service. The basic concerns of this team are:

1. ***Cost/benefit analysis.*** This should include manufacturing failures due to inferior supplies, slow-downs in manufacturing due to problems with supplies and services, and improvising due to inferior-quality supplies.
2. ***Specification development.*** The team will also review the percentage of product or service that presently meets specifications (the goal is 100%-adequate specifications). The specifications must be mutually agreed upon by the supplier and the customer. The team will assist in developing specifications that meet requirements and that can be verified and used for acceptance or rejection of the service or supply.
3. ***Specification development guidelines:***
   - **a.** What are the ultimate customers' expectations concerning the products being manufactured?
   - **b.** Quantify these concerns.

   c. Develop specifications that will satisfy these concerns.
   d. Determine if the printing process is capable of meeting these specifications.
   e. Determine acceptable variation.
   f. Use cause-and-effect analysis to determine those critical areas that can improve the process capability of an out-of-specification condition.
   g. Develop specifications for each of these variables.
   h. Develop a monitoring system that will assure that these variables are kept within specification.
   i. Continuously improve the process by analyzing, reducing, and eliminating variables.
4. ***Supplier measurement.*** Determine what you are going to measure in relationship to the quality of the product and the performance of the delivery. Define the percentage or variation that can be accepted and if that will comply to a process capability study (Cp). You may wish to do a process capability study to be sure that you can verify that what the customer needs is achievable. Look for simple measurement procedures. How easily can the data be obtained, and is it timely and relevant to the critical variables that will lead to customer satisfaction?

Applying the specification development process in printing. In this example the specification to be developed is the acceptability of photographs for conversion to halftone printing on coated paper.

1. Show various contrasting pictorials (low- to high-key examples of the same picture) to your customers for acceptance or rejection.
2. Determine if the pictures have critical elements that are more important to the customer.
3. Use a densitometer to measure the acceptable highlight and shadow tone to come up with a density range.
4. Set scanner curve to these densitometric readings and scan pictorial for a finished halftone with these specifications.
5. Run on press and measure with a densitometer to make sure the finished halftone print matches the density range required by the customer. If it does, this is your specification for this type of work.
6. Dot gain and other press variables may influence what specification you need on the process camera or scanner

in order to achieve the customer's expectations on the press.

7. After you have approved the halftone, through densitometric readings, measure 100 continuous press samples with a densitometer. This will give you the theoretical process capability. (Refer to Chapter 6 on how to do a process capability study). Beware that this capability is only for a very short duration of the 100 samples. To determine whether the process is in control, refer to Chapter 5 on how to set up and use a control chart.
8. Show the customer press samples, and let the customer determine which are acceptable and which are not.
9. Based upon what the customer accepts and what your process capability is, you can determine whether you can do this type of work without making process changes.
10. Determine, by the use of a cause-and-effect diagram, which critical areas will change the halftone and make it unacceptable to the customer. This process must include the people doing the job, their qualifications and their understanding of what has to be done, the materials you are using, the equipment and how it is maintained, the overall environment (which includes the mental frame of reference of the employees and the physical condition under which they work), and the methods by which specifications are communicated and maintained. If any of these change the halftone, it probably will end up out of specification.
11. Each of the cause-and-effect situations needs to have specifications drawn up so that it can be maintained within acceptable tolerances.
12. The press needs to be held to specification through preventive maintenance planning and execution. Oiling, cleaning, and blanket and roller checks and replacement will have to be done. Consistent attention through planned inspection of the press will need to be incorporated into the quality strategy of the company. All variables will need to be identified and monitored in a similar manner.
13. Develop flow charts of the present procedures and review them for process improvement strategies. This will enable your company to not only meet customer expectations but also allow for an opportunity to improve upon customer needs through improvements in tonal range and halftone reproduction.

Supplier selection. In developing the external supplier selection process, it is advisable to use the supplier survey strategy, which is discussed later in this chapter. It is very important to inform all employees of your company about the procedures that are used in selecting suppliers. In this way, everyone knows that his/her input is important and that the suppliers used by the printing company are the best available and have met the company's criteria for supplying the service or products needed.

Supplier improvement process. The objective of the supplier improvement process is to develop mutually beneficial partnerships. After grouping your supplies into such areas as paper, inks, and photographic film, you will need a vendor selection criteria procedure. The selection criteria for external suppliers are shown in the following table.

Selection Criteria for External Suppliers

- Product quality
- Delivery performance
- Vendor quality attitude
- Dollar volume
- Inventory lead time

Supplier awareness. It is extremely important for the SCP team to develop a methodology of how to make all suppliers aware of your company's total quality control process. A formal presentation should be made to all suppliers on what your company is embarking upon and explaining total quality concepts and how you view vendor partnerships. Top management and the team leader should make this presentation and introduce the team to the suppliers and explain what their role is.

This initial meeting should be followed by plant tours pointing out how their supplies are used and how important quality products, timely deliveries, service, and appropriate quantities are to your company's success. This portion of the plan should be carried out by team members.

The next step is for the team to survey your most important select suppliers to determine the appropriateness of purchasing from them and starting the development of a mutually acceptable continuous improvement partnership. This partnership must include a review of the communication systems that are in place and a means to ensure that the communication system is meaningful. Purchase requisition forms and delivery information must be part of the specification review and development process.

Phase II—The Supplier Survey

The objectives of the external supplier survey are to assure that all systems are adequate and that the vendor's process is capable of consistency. The object is to determine and eliminate anything in the present system that causes frequent and/or repetitive quality problems. The survey also assists in determining whether to place new business with the supplier and whether to develop a partnership that is long-lasting.

The following eleven items are the key areas that a printing company needs to consider when developing a supplier survey.

Supplier Surveys: Key Question Areas

- Management
- Financial
- Geographical
- Personnel
- Applied research
- Manufacturing
- Purchasing
- Inspection and testing
- Overall quality
- Customer focus
- Price

1. ***Management.*** The survey should look at vendor philosophy toward quality, organizational structure and if it is conducive to a quality product, how their vendors and personnel are indoctrinated toward a quality focus, and their overall commitment to quality. Some questions to raise as it relates to the management of your vendors include the following:
   a. Does the vendor have an operating philosophy that is conducive to a quality product?
   b. Does the mission statement emphasize quality?

- **c.** Does the vendor's organizational structure reflect a quest for quality?
- **d.** Does the vendor have a program for employees, their suppliers, and customers that explains the company's quality effort?
- **e.** Does the vendor espouse any of the quality gurus' philosophy? (Crosby, Deming, Juran, etc.)
- **f.** Does the supplier use the ISO 9000 standard or Baldrige criteria.

2. ***Financial.*** A general review of the financial standing of the vendor's business should look for such items as its philosophy of paying its bills, cash flow, and its equipment reinvestment policy. Some questions related to the financial situation of your vendors are:
   - **a.** Is the vendor financially stable?
   - **b.** How many new or relatively new pieces of equipment has the vendor purchased within the past five years?
   - **c.** Does the vendor offer discounts for early payment?
3. ***Geographical.*** Many companies are national and international in scope. Review where your vendors have their manufacturing sites. From whom do your suppliers purchase their supplies? Do these supplies come from various sources around the United States or other parts of the world? Are the vendors' manufacturing facilities located in areas that are known for natural disasters (hurricanes, tornadoes, earthquakes)? The answers to these questions will give you insights into product consistency and reliability.
4. ***Personnel.*** A review of how your vendor handles training and indoctrination of new employees will give some evidence of whether quality is a priority item of this organization. A low employee turnover rate is generally indicative of company harmony and stability. Some questions related to the personnel plan of your vendors include the following:
   - **a.** What is the labor turnover rate on average?
   - **b.** Is quality stressed as part of everyone's job?
   - **c.** How are basic job skills taught?
   - **d.** Do all personnel receive training in total quality control, including statistical process control?
5. ***Applied research.*** Companies that are involved in applied research are generally more open to process improvement. Some questions related to the applied research program of your vendors include the following:

a. Does the vendor review procedures and specifications for the products that it supplies to you?
b. How often does the vendor go about process improvement of its products?
c. Is the vendor capable and willing to work on new products?
d. Does the vendor keep up-to-date with technology and is the company committed to working on new quality methods?

6. ***Manufacturing.*** Companies that are in it "for the long haul" generally have a program of updating manufacturing equipment. They are always looking for better equipment and attend equipment manufacturers' shows to review new and emerging equipment. Some vendors have machine shops in which modifications to existing equipment are constantly being made. Also, the manner in which maintenance and housekeeping are performed is an indication of the quality effort of the vendor. Some questions related to the manufacturing facility of your vendors include the following:
   a. Does the supplier have modern equipment or equipment that is capable of doing quality work?
   b. Does the vendor have a preventive maintenance program?
   c. Is the process stable? Does the vendor have evidence of this stability?
   d. Does the vendor do capability studies?
   e. Does the vendor use statistical process control to maintain quality levels?
   f. Does the vendor have adequate production capability?
7. ***Purchasing.*** How your vendor purchases supplies and the efforts that your vendor makes to have specifications, order quantities, and delivery dates adhered to by its suppliers indicate your vendor's ability to manufacture a quality product at a competitive price for you. Some questions related to the purchasing policy of your vendors include the following:
   a. Is the vendor open to working with you in developing product specifications, communications, and service requirements?
   b. Does the vendor work with a single-source supplier?
   c. Is the vendor a single-source supplier to any of its customers?

d. What is the structure of your vendor's single-source arrangement?
e. Does the vendor desire to be a single-source supplier?

8. ***Inspection and testing.*** The laboratory facilities, testing equipment, and procedures of calibrating and maintaining laboratory equipment are all indicators of the quality auditing system and process improvement commitment of your vendor. Some questions related to the inspection and testing environment of your vendors include the following:
   a. How does the vendor monitor incoming supplies?
   b. Does the vendor have an inspection and/or testing laboratory?
   c. Is the vendor willing to share inspection and testing information?
   d. Is the laboratory, its equipment, personnel, and procedures adequate for verification of the quality product you desire?
   e. What are the sampling techniques used by the vendor?
   f. How is the testing and evaluation data utilized?
   g. Does the vendor have calibration procedures that are nationally recognized and controlled?
9. ***Overall quality.*** Your survey must look at the vendor's quality in terms of what procedures it has that ensure a quality product to you. Vendors must pay attention to their quality effort. You are interested in quality performance, which requires a commitment to the process of quality on the part of your vendor. Overall quality includes the quality of your vendors' suppliers and customers. Some questions related to the overall quality of your vendors include the following:
   a. Does every department of the vendor's organization use a quality-based process and how much can be quantified? Look for flow charts, cause-and-effect diagrams, and control charts.
   b. Does the vendor have a corrective action plan?
   c. How does the vendor deal with nonconforming supplies?
   d. How does the vendor deal with nonconforming products?
   e. Will the vendor supply samples of its product with shipment?
   f. Will the vendor work with you in developing a quality strategy?

g. Will the vendor supply statistical evidence of conformance to quality?
h. What technical services will the vendor supply?
i. What training services will the vendor supply?

10. ***Customer focus.*** The vendor must view you, the customer, as the ultimate consumer. Your vendor must have a strategy of working with customers to develop technical specifications, communication systems, service, and training aids.
    a. Does the vendor have a follow-up procedure to determine how the product or service being supplied is perceived by your company?
    b. Do service and technical support staff visit on a regular basis to determine your needs?
    c. Does the vendor know who your customers are and what you must do to satisfy them?
    d. Has your vendor made an effort to get to know your problems as they relate to their products?
11. ***Price.*** Price can only be considered in view of the total price the supply will cost. If the vendor's quality can not be assured or the process is not adequate for long-range conformance to specifications of the product, then rejects, rework, higher inventories, financial readjustments, and unhappy customers will be the outcome. A printer must balance service, quality effort, delivery, partnering, and delivery dates when choosing a supplier. Some questions related to price considerations of your vendor's product include the following:
    a. Is the vendor's price competitive?
    b. If the vendor is higher or lower, why?

Phase III—Supplier Assistance

Supplier corrective action plan. The objective of a supplier corrective action program is to open up two-way communication for the handling of discrepancies in shipments. It allows the supplier to identify and eliminate root causes of these discrepancies and provides a follow-up mechanism to assure that the corrective action is effective.

Supplier corrective action form. The objective of the supplier corrective action form is to verify that the printer is not at fault for the quality discrepancy. Both the printer and the supplier must know that the purpose of this form and procedure is not to place blame but to determine responsibility and get to the root cause and eliminate this quality defect.

External Supplier Quality Information Request

Purposes of This Questionnaire:

1. To develop closer relations with our suppliers.
2. To work toward preferred supplier status.
3. To develop partnerships.

General Overview Issues:

1. Please attach an organizational chart.
2. Please attach your mission statement.
3. How many manufacturing plants do you have?
4. Where are your manufacturing plants located?
5. What is your total number of employees?
6. List your ten top customers and the percentage of business they do with you.
7. List the equipment you have purchased in the last five years.
8. Describe your preventive maintenance program.
9. What is your educational policy toward employees?
10. How much do you spend on training and retraining of employees?
11. Do you have a customer education program? Please describe.
12. How many technical assistance people do you have?
13. What professional organizations does your company and its employees belong to?
14. What is your company's or its employees' involvement with these organizations?

Quality-Related Issues:

1. Describe your quality philosophy.
2. Attach an organizational chart of your quality assurance department.
3. Explain the educational background of your quality assurance, sales, and technical support staff.
4. Do you have a research facility?
5. Do you use statistical process control?
   a. If yes, which SPC tools do you use?
   b. Supply samples of how you use SPC.
6. Describe the quality control and testing equipment you have.
7. What is your procedure for quality control in manufacturing?
8. How do you evaluate incoming raw materials?
9. List the equipment you plan to purchase in the next two years.
10. How do you report materials failure?
11. What do you do with failed materials?
12. How many employees have quality assurance as their primary function?

No one wins with mistakes; it is therefore important to develop a corrective action plan that will eliminate mistakes.

The procedures for using this form include the analysis of defects to provide feedback to the supplier. This form should finalize the corrective action and document the occurrence of deviation from the specification.

Corrective Action Form

Problem ____

Reported by ____________________________ Date ______________

Specifications of all component parts _____________________

____

Verification of specifications that were met _________________

____

Deviations in specifications ______________________________

____

Corrective action taken _________________________________

____

Process improvement taken or needed ____________________

____

Signature ______________________________ Date ______________

Supplier assistance activity. The supplier survey is used as the planning tool for the supplier assistance action plan. At the beginning of your materials specification development process, it is a good idea to have frequently planned meetings with your suppliers to review the survey and develop an action plan that includes a review of specifications. This is usually a mutual training session where your supplier learns about your needs and you learn about the supplier's capabilities.

Phase IV—Supplier Certification

The objective of a supplier certification program is to develop an understanding, with your suppliers, that your printing company and your suppliers have adequate systems, procedures, and controls in place to begin to work toward 100% defect-free materials. ISO 9000 is one method of certifying that a system is in place that should ensure consistency. However, ISO 9000 does not ensure quality, and therefore a complete understanding of specifications is important between you and your supplier.

Certification of suppliers means that the supplier has demonstrated process capability to ensure constant quality on a timely basis. Certification supplants normal incoming testing and inspection. Auditing on a prescribed basis can verify that all systems are in place. This can take place quarterly, semi-annually, or yearly. The certificate denotes reliance upon the supplier for a consistent quality product.

Certification requirements. The requirements for certification must include survey completion, performance feedback concerning quality and delivery of supplies, adequate systems that will ensure consistent quality, demonstrated process capability, and appropriate quality control procedures to maintain process capability.

Supplier recognition. It is advisable to develop a supplier recognition policy. Your quality printing depends upon your suppliers' efforts in getting their products to you in a quality manner. A supplier recognition program should be documented, measurable, and communicated to all of your suppliers.

Recognition of consistent quality can be rewarded in many ways. Parties, luncheons, dinners, plaques, and letters of thanks are just some of the ways that you can recognize and promote quality in your supplier's organization.

Supplier councils. Supplier councils offer a printing company an opportunity to have its preferred suppliers assist in developing strategies for process improvement within the printing environment. Rarely does the paper supplier visit and discuss avenues of progress with the color separator, press manufacturer, inkmaker, and plate manufacturer. Generally, they never see each other and when they do it is to find the "guilty party." If a telephone conversation takes place between them, it is because everyone is looking for someone to blame for a quality problem or they may be adversaries in legal cases.

Planned supplier council meetings, with the appropriate team involved, will place both printer and supplier on the same team, one of quality improvement through agreed-upon specifications and a mutual commitment to the final customer.

Working through Predictable Situations

Vendor resistance to supplier certification. The supplier certification process means accountability and a change of procedures for some of your vendors. These are not always welcome. Your suppliers may see the goals of the supplier certification process as unrealistic if not totally impossible. This can be expected and needs to be dealt with in a realistic and understanding manner. Suppliers may feel threatened by your requests for information about their companies. You must assure them that your reasons are to develop ongoing partnerships and these questions reaffirm your respect for their products and services.

To keep your business, they may start but will falter if they are doing it to humor you and your supplier certification process. They will find all types of reasons why it can not work. You must remain consistent in the effort and not allow distractions to halt the process.

Procedures for overcoming supplier resistance to the certification process. The following are useful tools that should be used as part of a company's strategy in developing the four phases of the supplier certification process. The following methodology applies to internal and external suppliers.

1. Emphasize benefits of internal and external partnerships. Work toward an understanding of openness so that both vendor and customer can visualize the benefits of the certification process.
2. Involve the supplier in the development of specifications and procedures. Where this can not be done, involve the supplier in understanding why the specification is critical to the printing process.
3. Educate and train suppliers in the methods and procedures of the four phases of the supplier certification process and your overall thrust toward a TQM company.
4. Be totally open about what you are doing and how you intend to implement a supplier certification process and encourage input as to the implementation process.
5. Always be willing to share information and be open to educating and training supplier personnel as part of your ongoing insistence on the supplier certification process.
6. Make sure that all lines of communication are open and that your supplier is kept totally informed as to your needs and expectations. Develop procedures for dealing with misunderstandings. Be open to early detection of nonconformance to specifications.
7. Continually reinforce that the supplier is part of the quality team. The supplier needs to be encouraged to view its products as critical to your customers.

8. It is a hard pill to swallow, but suppliers need to understand that they are replaceable. Suppliers need to know that you wish to develop partnerships with them; however, you must insist upon the development of SCP. If the supplier refuses, then that supplier must know that you will search for a supplier who will work with the process.
9. SCP is a long-term commitment, and you and your supplier must be aware that the outcome will help both of you and it is worth the initial effort.
10. The advantages of SCP must always be reinforced. When the initial thrill of newness wears off, the hard work will still remain. We must keep in mind the advantages that will be the outcome from SCP.
11. Develop a "Vendor Day" or some other means of recognizing your preferred suppliers. Awards, including information about your suppliers in your newsletter, promoting SCP and what you and your suppliers have accomplished, and writing letters of congratulations are all means of keeping suppliers involved and enthused with SCP.
12. Certificates of preferred supplier status give suppliers a sense of accomplishment and the needed recognition to continue their SCP efforts. Be aware that these certificates need to be truly earned and suppliers need to know what they must do to win these awards. Giving certificates out frivolously will dilute your efforts toward conformance to specifications.

Finding the time. SCP is one of the initial steps in the development process of total quality management of your company. Time has to be allocated to this part of the plan. Keep in mind that you must have a plan which includes time, people, and delegation of responsibilities. Make sure that your outcomes are clearly focused. Do not get your vendors involved until you have a clear focus of how you intend to implement SCP. Set up a time schedule for the four phases, which includes the money necessary to free up your employees to do the task. Do not make SCP additive to a normal workday.

Quick fix. SCP is not a quick fix. It will take time and energy to accomplish partnership relations with your suppliers. Keep in mind what you have to lose by not developing an SCP plan.

Training. Make sure your team is trained in total quality control procedures. Allow time and financial support to the training process of the SCP team. Do not overlook the training that will be required to train your suppliers in what your company is embarking upon.

Communications. Be sure to have a communication strategy well developed before you embark upon SCP. Internal suppliers and customers need to know why SCP is being done and the benefits to all after SCP is implemented.

Keep at it—do not get discouraged. Expect pitfalls. Expect that you will uncover a lack of specifications, monitoring systems, poor communication delays, and "make-do" mentality within your organization. Do not give up! It will take time and persistence to develop a complete set of specifications for all of your operations. Identify the most important items that will give you recognition that specifications are worthwhile. Work toward developing specifications for all items within your printing company.

Customer-recommended or -supplied materials. In the printing industry, it is a common occurrence for the customer to recommend or supply materials for their printing job. The general stand that printers have taken to this occurrence is "to grin and bear it." This is the wrong procedure. Customers often use their buying power to leverage a given price for their supplies. There is nothing wrong with this. A low price for supplies is a worthy achievement for the customer. It is assumed that the bottom-line cost of the printing job will be lower. It is important that all specifications for that material be agreed upon by you, your customer, and the supplier. It is your duty and right to work with the customer to develop specifications that work for the printing job and are feasible for you. A procedure must be developed that encompasses a statistically sound process for these customer-purchased materials. If the supplied materials have no specifications or are out of specification, the customer must know the chances of not getting a quality product if you must "make do" with these materials.

Summary

In this chapter, we have shown how the systematic approach of a supplier certification process can be used to reach the goal of consistent quality supplies from key suppliers. Supplier certification implies a partnership in which there is trust that the supplier will continually work toward meeting not only the printer's current needs but also remain vigilant to changing needs. Organizations have to realize that they have to deal not only with external suppliers, but that different departments within the printing company are in actuality suppliers and customers to other departments.

12 Practical Applications of the Baldrige Award Criteria

Applications to the Printing Industry

The work done by many experts involved in government, business, and education has led to the development of a quality strategy known as the Malcolm Baldrige National Quality Award (MBNQA). Even though this chapter describes how to officially apply for this prestigious award, it is suggested that most printers only use the methodology to analyze their strengths and weaknesses, but not necessarily apply to win this award. Some companies use the MBNQA application process as a means of getting expert feedback. Others take a long-range view and apply with the thought that they will utilize the feedback to improve their quality process each year until they can win. If used properly, the MBNQA analysis process can prove worthwhile in helping a printing company to understand what is considered a quality organization and where they may be weak or strong in relationship to this criteria and give the organization a quality improvement direction.

Historical Background and Purpose

The Malcolm Baldrige National Quality Award was signed into law on August 20, 1987. The award is named for Malcolm Baldrige who served as Secretary of Commerce from 1981 to 1987, when he died in a rodeo accident. Public Law 100-107 led to the creation of the Foundation for the Malcolm Baldrige National Quality Award, a private-public corporation that is the financial backing for this award.

The basic tenets of the MBNQA are that the United States is being challenged by foreign competition in process quality and that our competitors have improved their productivity to a greater degree than we have in the last two decades. Poor quality can cost companies as much as 30% of sales revenue while improvement in quality and services can lead to improved productivity, lower costs, and increased profits.

To compete effectively in the global marketplace and for the well-being of our nation's economy, strategic planning of quality through quality improvement programs will be required. Management must understand and lead the work force in a personal commitment to quality and the use of quantitative analysis tools such as statistical process control. Quality improvement is a concept that is applicable to all segments of our economy from large to small printing companies as well as those in the public sector. Successful quality improvement programs must be management-led and customer-focused.

The MBNQA was set up to encourage American companies to improve quality and productivity by recognizing those companies which implemented and maintained quality improvement processes that lead to a competitive edge and increasing profits. Using the recognition process, those companies that have improved the quality of their goods and services provide a benchmark for other companies. The MBNQA established guidelines and criteria that can be used by all segments of the American economy to evaluate and improve their quality efforts. Each year the MBNQA guidelines are reviewed and modified, where necessary, to improve the process of identifying and rewarding those American companies that are determined to improve their quality through a significant process improvement strategy.

How the Award Is Set Up

The Foundation of the MBNQA was created with its main objective to raise funds to permanently endow the mission of the award.

The Department of Commerce is assigned the responsibility of the award, which it has assigned to the National Institute of Standards and Technology (NIST) to manage. In turn, the American Society for Quality Control (ASQC) has a contract to assist in the administration of the award program. NIST conducts basic and applied research in the physical sciences and engineering and develops measurement techniques, test methods, and standards. ASQC has the mission of identifying, communicating, and promoting the use of quality principles to facilitate customer satisfaction through continuous improvement.

The Secretary of Commerce appoints a Board of Overseers to evaluate all aspects of the MBNQA content, criteria, and process. The board is also responsible for assessing how the MBNQA is serving the nation's interest. The board is a recommending body to the Secretary of Commerce and the director of NIST.

The Board of Examiners evaluates award applications, writes feedback papers, and recommends award winners to the director of NIST. To date, there are over 270 examiners, mostly from the private sector. All examiners must be trained in the MBNQA process.

There is a maximum of six awards given annually, with a maximum of two each in the categories of manufacturing, service, and small business (less then 500 employees) company. The process is set up in four stages: (1) at least five examiners review the application, (2) a consensus review of how well the company scored, (3) a site visit, if warranted, and (4) the final review. Each applicant receives a feedback report at the end of the process.

The cost of the application process includes a $50 nonrefundable eligibility fee, for manufacturing or service companies a $4,000 fee, for small business a $1,200 fee, and if you have more then one type of business (e.g., printing and publishing or broadcasting), an extra $1,500 for a supplemental form.

All award winners must share information on their success. However, proprietary information does not have to be shared. This is generally done at the annual Quest for Excellence Conference.

The MBNQA Evaluation Categories, Items, and Points

The examination criteria for the Malcolm Baldrige National Quality Award has seven categories that are broken into twenty-four items with a total value of 1,000 points. The table on the next page lists these categories and items and includes the point values for each.

Evaluation by Approach, Deployment, and Results

The twenty-four items and the fifty-four addressable areas that comprise the examination criteria are evaluated in respect to approach, deployment, and results of each item. The approach is to determine how appropriate are the methods, tools, and techniques being used and whether these are systematic and consistent in the entire organization. The company must use information that is objective and quantifiable. Deployment is how a printing company applies the approaches of the items being evaluated throughout all relevant areas in the company. The assessment team looks for appropriateness and effectiveness in key processes and how this relates to the development and delivery of the product and services. Approach also relates to how the company

Malcolm Baldrige National Quality Award 1995 Award Examination Criteria

Categories and Items		Point Values
Leadership		**90**
Senior Executive Leadership	45	
Leadership System and Organization	25	
Public Responsibility and Corporate Citizenship	20	
Information and Analysis		**75**
Management of Information and Data	20	
Competitive Comparisons and Benchmarking	15	
Analysis and Uses of Company-Level Data	40	
Strategic Quality Planning		**55**
Strategic Development	35	
Strategic Deployment	20	
Human Resource Development and Management		**140**
Human Resource Planning and Evaluation	20	
High Performance Work Systems	45	
Employee Education, Training, and Development	50	
Employee Well-Being and Satisfaction	25	
Process Management		**140**
Design and Introduction of Products and Services	40	
Process Management: Product and Service Production and Delivery	40	
Process Management: Support Services	30	
Management of Supplier Performance	30	
Business Results		**250**
Product and Service Quality Results	75	
Company Operational and Financial Results	130	
Supplier Performance Results	45	
Customer Focus and Satisfaction		**250**
Customer and Market Knowledge	30	
Customer Relationship Management	30	
Customer Satisfaction Determination	30	
Customer Satisfaction Results	100	
Customer Satisfaction Comparison	60	
	Total Points	**1,000**

interacts with customers, employees, and the general public. Results include present performance levels and quantitative proof that positive changes are taking place. Also analyzed, under the results category, would be the printing company's rate of improvement and whether this improvement is being sustained throughout the organization. Performance levels also are related to appropriate comparisons and/or benchmarks. Some of the twenty-four items require that only approach and deployment be evaluated, while others require results alone.

Scoring of approach, deployment, and results is done on a 0–100 scale. Approach and deployment are treated together, while results are evaluated separately. Zero indicates that there is no systematic approach or information available and the results are poor. A score of 10–30% means that there is the beginnings of a systematic approach and the company is in the early stages of development with major gaps in the approach and deployment of the specific item being evaluated. There may be good performance in some areas, but basically the results are spotty in important areas. A score of 40–60% means that there is a fairly sound approach, but in some areas there is little deployment. The results are still spotty, but improvement trends are good. A rating of 70–90% means that there is a sound approach in all areas; however, the deployment in some areas may vary. Results are very good, with good to excellent applications in most relevant areas of printing. To receive a rating of 100% requires a sound and systematic approach to all items where each item is fully deployed. The results are excellent, and there is strong evidence that there is benchmark leadership in many areas.

Defining the Items of the MBNQA

Leadership. The ***Leadership*** category examines how top management provides personal leadership in creating and maintaining customer focus, understandable values and expectations, and a system that promotes performance excellence. How these values are expressed in terms of public responsibility are also examined.

"Senior Executive Leadership" is reviewed as to how senior executives (top management and those directly reporting to them) lead and direct in developing the printing company's competitiveness, performance, and capabilities. The evaluators review how top executives set direction through the entire company. The evaluation is based upon performance excellence, values, and expectations. These items can be evaluated with the use of board of director reports, peer evaluation, and employee surveys.

"Leadership System and Organization" examines how top management focuses on customers and high-performance objectives. A review is made of how effectively the organization communicates and reinforces its values, expectations, and directions and how work performance is evaluated to improve the process.

"Public Responsibility and Corporate Citizenship" analysis deals with legal issues such as OSHA, MSDS, and VOC regulations. Does the printing company look ahead to anticipate public concerns? Is the company involved in civic responsibilities and the promotion of legal and ethical behavior both within and outside the company?

The above three items are all evaluated by approach and deployment.

Information and Analysis. The ***Information and Analysis*** category surveys management effectiveness in the use of data and information to support marketplace success based upon excellence in customer-driven performance.

"Management of Information and Data" deals with how information and data is selected and used for improvement purposes. The data should focus on key business drivers and is evaluated by how well this data is used for better business practices.

"Competitive Comparisons and Benchmarking" surveys the criteria used for determining what to benchmark both from within and outside the organization. The benchmarks should be customer-related, product or service quality, internal operations to include internal business functions, support services, employee relations, and supplier performance.

"Analysis and Use of Company-Level Data" is a review of how a printing company uses data from all parts of the organization to support reviews, business decisions, and planning. The information must be used to understand customers and markets, company capabilities and performance, and competitive performance. An evaluation is also made of how this information is used for setting priorities and improvement activities.

The above three items are all evaluated by approach and deployment.

Strategic Planning. The ***Strategic Planning*** category investigates how a printing company develops and implements its strategic directions and how it determines the key elements of this strategy.

"Strategy Development" examines how a company plans for overall performance in the short and longer term. The relationship of customer requirements, competitive position, company capabilities, and the risk involved are considered.

"Strategy Deployment" summarizes the key business drivers and how they are translated into an action plan. Consideration is given to cycle time improvements, inventories, work-in-progress, inspection, downtime, changeover time, set-up time, and any other examples of the use of resources and materials. MBNQA examiners will take notice of any differences between any short-term and long-term plans.

The above two items are both evaluated by approach and deployment.

Human Resource Development and Management. The ***Human Resource Development and Management*** category surveys how employees are developed to their full potential as related to the organizational objectives. This criteria also examines how the work force is enabled to fully participate in excellent performance for personal and organizational growth.

"Human Resource Planning and Evaluation" involves translating strategic planning into specific human resource plans. Some of the items that are reviewed are how the company deals with work design flexibility, employee development, recognition, and recruitment.

"High Performance Work Systems" investigates how job design promotes high performance. The investigation includes an analysis of self-directed responsibility, effective communications, and appropriate team efforts.

"Employee Education, Training, and Development" evaluates how employee education and training build employee capabilities. This area addresses the knowledge and skills necessary to meet individual work objectives. A review of how education and training is evaluated and improved are also important factors that are considered.

"Employee Well-Being and Satisfaction" analyzes how the company maintains a safe and healthful workplace. This study includes a review of health, safety, and ergonomic improvement activities. Areas that may be reviewed include turnover rate, absenteeism, strikes, grievances, and worker compensation claims.

All four of the items in this category require an approach and deployment.

Process Management. The ***Process Management*** category examines customer-focused design elements, product and service delivery, research and development, and support services. High performance is the key component as it relates to how processes are designed, effectively managed, and improved.

"Design and Introduction of Products and Services" surveys how new and/or modified products, services, and delivery systems meet quality and production requirements. A review of how these elements complement customer requirements and how these are evaluated and improved are also considered.

"Process Management: Product and Service Production and Delivery" defines how central elements of a company's product and service production/delivery processes are managed. The basis of this analysis is to ensure that quality and operational performance are continuously improved.

"Process Management Support Services" reviews how these services are designed and managed to support the focus of the organization. Some of these services are estimating, customer service, accounting, and facilities management.

"Management of Supplier Performance" investigates how the company ensures that suppliers meet performance requirements. A review of key performance requirements is made along with how these are communicated to suppliers along with how performance levels are ascertained and fed back to the supplier. Feedback systems are reviewed not only for external but for internal suppliers as well.

All four of the items in this category require an approach and deployment.

Business Results. The ***Business Results*** criteria reviews the performance of the company and how it is improving in its primary business areas. Product and service quality, productivity, and financial performance are compared with the competitors.

"Product and Service Quality Results" are evaluated on how product and service quality have improved. Customer satisfaction and customer retention are carefully scrutinized. The use of graphs and tables are used in order to develop comparative data.

"Company Operational and Financial Results" summarizes the improvement efforts employed by the company by analyzing key elements of operational and financial performance. The various indicators can include productivity, cycle time, cost reductions, absenteeism, turnover, and environmental improvements.

"Supplier Performance Results" deals with the results of supplier performance improvement. A review is made of supplier partnerships, reducing the number of suppliers used, along with appropriate industry benchmarks.

All three of the items in this category require the use of results as the focus of the investigation.

Customer Focus and Satisfaction. The ***Customer Focus and Satisfaction*** category evaluates the printing company's system for maintaining and building customer relationships. A review is made of customer satisfaction, retention, market share, and how these items also relate to the competition.

"Customer and Market Knowledge" looks at how the company determines not only short-term but also longer-term requirements and expectations of its customers. A review is made of how the company has developed strategies to anticipate customer needs. The investigation looks at such items as how complaints and how gains and losses in customers are used by the company to support customer and market knowledge.

"Customer Relationship Management" surveys how the company manages its efforts in responses and follow up with customers. The effort is reviewed with the main intent of determining how the company uses these strategies to preserve and build relationships with its customers. How the use of handling complaints, the elimination of the causes of complaints, and the recovery of customer confidence after complaints is of prime importance in this area of investigation.

"Customer Satisfaction Determination" involves how the organization determines customer satisfaction and also how it relates to the competition. Information as to customer reorders, referrals, and how all of these quantifiers are used to improve customer satisfaction throughout the company are also reviewed.

"Customer Satisfaction Results" identifies and summarizes the printing company's customer satisfaction and dissatisfaction results. The results are segmented by customer type and address the most important indicators of the organization's products and services.

"Customer Satisfaction Comparison" compares the printing company's customer satisfaction results with that of the competitors. Some of these indicators include gains and losses in customers and are segmented by customer group where appropriate.

"Customer and Market Knowledge," "Customer Relationship Management," "Customer Satisfaction Determination," and "Customer Satisfaction Determination" require an approach and deployment while "Customer Satisfaction Results" and "Customer Satisfaction Comparison" are evaluated by results.

Benchmarking the Printing Company with MBNQA Criteria

The assessment form on page 291 was developed to mirror the MBNQA. It can be used by a printer to do a self-evaluation on how the company compares to some of the basic criteria used by the Malcolm Baldrige National Quality Award. This benchmark can then be used to improve the processes that are considered critical in developing a quality program. This program must be continuous and built on sound process improvement strategies. It is suggested that more than one person do this evaluation in the printing company. Compare these perceptions to determine where the basic strengths and weaknesses arise and use brainstorming techniques to determine a plan for process improvement.

Instructions for using the self-assessment form. The ten key concepts shown on the self-assessment form are basic to the Malcolm Baldrige National Quality Award. Each concept has three categories to use for evaluation. These categories are approach, deployment, and result. Place a check mark in each category that you perceive your printing company is doing well in. Check marks must go from left to right. If there is no approach, there can be no deployment. If there is no deployment, there can not be any results. When the form has been completed, add up the check marks and review the score to determine where you feel your printing company now stands as compared to the Malcolm Baldrige National Quality Award criteria.

Definition of key terms used with the assessment form. The following are definitions for the terms "approach," "deployment," and "results" used with the assessment form:

Printing Company Quality Self-Evaluation Form

(Generally based on the Malcolm Baldrige National Quality Award Assessment Process)

No.	Quality Concept	Approach	Deployment	Results
1	Is quality defined by your customer?	______	______	______
2	Does senior management create quality values and build these values into how the printing company operates?	______	______	______
3	Is quality judged on well-designed and well-executed system?	______	______	______
4	Is continuous improvement emphasized in all areas of the organization?	______	______	______
5	Does your company have goals, strategic plans, and operational plans to achieve quality?	______	______	______
6	Does your company have process improvement plans for shortening response time in all operations?	______	______	______
7	Are the operations and decisions of the printing company based upon facts and data?	______	______	______
8	Are all employees trained and utilizing quality procedures in their work?	______	______	______
9	Is quality designed into each job with defect and error prevention a major part of the quality system?	______	______	______
10	Are quality requirements and specifications communicated to suppliers and are supplier partnerships emphasized with the goal of continually improving these elements?	______	______	______

Total Score: __________

- ***Approach:*** There is a fundamental understanding of the concept that is agreed upon by key employees and is customer-based, systematic, and prevention-oriented.
- ***Deployment:*** The approach is effectively implemented in the key areas of the printing business.
- ***Results:*** There is quantitative proof that positive results are taking place from the approach and how it is being used.

Explanation of Score

Score*	Meaning
0–5	No evidence of any quality effort. Essentially no attention is given to quality.
6–9	There is a slight effort given to any quality concept. Quality is considered low priority.
10–13	Some effort in quality concepts. Nothing is outstanding with little integration of quality efforts.
14–17	The company has a good prevention-based quality effort; however, many areas lack consistency of a quality effort. There is a significant need for more deployment and results to demonstrate process improvement.
18–21	There is evidence of significant effort in most concepts and outstanding in several. Not all areas of the organization are at the same level of quality, but these areas can clearly be improved.
22–26	All areas of the company have good results, and some are even outstanding. All areas are integrated with good to excellent results. You should be a printing industry leader.
27–30	There is outstanding effort in all areas of the company. You have proven that all areas are integrated and have sustained results to prove it. Your printing company is a national benchmark.

*To determine your score based on the 100-point allocation that the MBNQA uses, multiply your score by 3.3. As an example, if you add up your evaluation of the organization and it totals nineteen check marks, multiply 19 by 3.3, which will equal 62.7.

Summary

In this chapter we have discussed the use of the Malcolm Baldrige Award criteria as a method of benchmarking your printing company's quality progress in relationship to what is considered an outstanding American company. The use of the MBNQA is an all-inclusive benchmark focusing on your customer and how customer satisfaction is achieved. This benchmark can then be used as a means of process improvement for your company.

13 Using ISO 9000 to Improve Print Quality

Introduction to ISO 9000

Over the past several years, we in the printing industry have been hearing about a new standard coming from Europe called ISO 9000. It has been said that, to do business in the unified market of Europe, we would have to become certified to this standard. Many printers let out a big sigh of relief—since most of us don't print for the European market—and have gone about doing business as if they hadn't heard anything at all. If that is the case with your company, it would be a good idea to look again because ISO 9000, while not representing much of a threat to most printers in the sense of further regulation, represents a significant opportunity for jump-starting your quality improvement process. In some printing markets the standard will, in fact, become a prerequisite to doing business. These markets will probably include pharmaceutical printing, industrial labels, package printing, printing for the automotive industry, technical documentation, and any type of printing for the computer industry. If you print for companies that do business in Europe or for other international markets, expect to be asked about your compliance with ISO 9000. In other markets, such as commercial, book printing, and publications printing, the standard can provide a company with marketing advantages, even if not a prerequisite. Finally, in other companies—actually in all companies—pursuing the standard and going through the process of getting certified, can provide an excellent means for jump-starting your quality improvement process. It should be noted that the ISO 9000 standards are scheduled to be revised every five years. Thus, if the reader has either copies of the standards or books and articles discussing ISO 9000, be sure to check that you have the ANSI/ASQC Q9000-1994 revision.

In many cases of quality improvement implementation, there is a lack of direction and a weakness in the underlying quality foundation. This is where ISO 9000 might fit in for you. In this chapter, we discuss the following topics:

1. What is the ISO 9000 standards series?
2. How can ISO 9000 be implemented?
3. Types and purposes of quality audits.
4. What are the costs and benefits involved?

What Is the ISO 9000 Series of Standards?

Background. ISO 9000 is a generic system that specifies in very broad terms the necessary components of a quality system. ISO 9000 is not specific to printing, or any single industry for that matter. Rather, it details a list of standards that encompass the quality function for all industries. ISO 9000 was originally published in 1987 by the International Organization for Standardization in Geneva, Switzerland. The standards were written by an international group of quality experts and practitioners including input from the United States. The series was developed as part of the European push toward a fully integrated common market. As part of this movement, it was recognized that for the common market to exist there would have to be assurance of quality across borders. The eleven countries of the European Community (EC) thus adopted ISO 9000 as the basic minimum quality system for selling certain products (such as electronic components and construction equipment) in the EC. But ISO 9000 has blossomed into much more than its intended original purpose. Nowadays, virtually anybody who wants to sell a product in Europe or sell a product to a European company must take a serious look at ISO 9000 registration.

Most countries have at this point adopted the ISO 9000 standards for use within their own borders. This has led to some confusion, as not all adaptations are 100% compatible with each other. For example, the British version is called BS 5750 and was the foundation of the ISO 9000 series. The American version, which was developed by the American Society for Quality Control and the American National Standards Institute is known as the Q90 series. If a company in the United States becomes registered as conforming to the Q90 series, it must still become registered under the British version or be registered simultaneously under both standards by a registrar, such as Underwriter Laboratories, that is qualified to do so in conjunction with the British Standards Institution. For the most part, however, the American Q90 standard is fully compatible with ISO 9000—if you become registered under one, it is relatively easy to become registered under the other, if it should become necessary to do so. The major difference is in the use of American English and spelling.

The registration function is performed by an organization known as a registrar. These are mostly private companies whose purpose is to perform third-party audits and verify that a company is in compliance with ISO 9000. These groups are registered with a group known as the Registrar Accreditation Board (RAB) in the U.S.A. (Other such accreditation bodies are the NAACB in England or the RVC in Denmark.) Working for the registrar are individual auditors, who make plant audits and decide whether the company can be registered.

The system. Actually, ISO 9000 is a series of five documents that work together as a complete quality system. These documents and their content are as follows:

- ISO 9000 (Q90) is the basic set of guidelines for the selection and use of management and quality assurance standards. It is a statement of purpose and a set of definitions that serves an advisory function. It suggests whether your company should pursue ISO 9001, 9002, or 9003 registration.
- ISO 9001, 9002, and 9003 (Q91, 92, and 93) are the actual standards in the system to which a company becomes registered; each of the three represents different models for a quality system depending upon the type of business a firm is in. They differ in scope and are generally selected for use based on the type of business the company is in. The most comprehensive is ISO 9001, covering twenty different components of a quality system. These range from the responsibility of management in setting quality policy and defining quality responsibilities to areas such as purchasing processes, training procedures, and corrective action methodologies. ISO 9002 is less comprehensive, omitting the necessity to look at design control (research and development). Most printing companies will use ISO 9002. ISO 9003 is least comprehensive, additionally omitting functions such as purchasing, design control, process control, and servicing.
- ISO 9004 (Q94) is a generic template of the various elements of a quality management and assurance system. It covers such items as economics, quality in procurement, quality in marketing, and the use of statistical methods. Essentially, ISO 9004 is a guideline for implementing and auditing the total quality process. If your purpose in looking

at ISO 9000 is to find a model for your quality system to be built upon, then ISO 9004 is an excellent source. ISO 9004 explains basic issues that you need to consider when implementing your ISO strategy. However, a company cannot be certified (registered) to ISO 9004, only to ISO 9001, 9002, or 9003.

A company wishing to pursue ISO 9000 registration would select one of the three systems applicable (9001–3) to its individual situation. In the printing environment, given the importance of the purchasing and production functions, we would recommend selecting at least ISO 9002. In fact, less than 2% of all registrations are to the 9003 standard, which is primarily for inspection and testing processes. If you are in manufacturing, as printers are, you will want to look for a higher standard. If you are heavily involved in research and/or product design (both aesthetic and functional as in packaging or industrial labels), you would be well suited to choose the ISO 9001 standard as your preferred template. The differentiating factor between selecting ISO 9001 or 9002 is the extent to which the printer is involved in the design of the product. Many printers in the commercial market provide essentially a commodity service for their customers. If this is the case for your company, the 9002 standard is appropriate. The key to selecting the appropriate standard is in looking at the type of business the firm is in. If it is a manufacturing-intensive firm, without extensive research and development (design), as most printers are, ISO 9002 is the appropriate standard.

The chart on the next page shows the contents of each of the three standards and which contents aren't covered as you move from 9001 to 9003:

While all of this might sound complex, in essence the ISO 9000 standard is a very simple approach to developing a quality system. It is based on the principle of the standardized operating procedure and auditing to that procedure. In essence, in order to become ISO 9000 certified, a company must do the following three things:

1. Document what you do, especially if there is an effect on product quality—this means that you must write down exactly how you take an order, make a plate, or run a press as it relates to the quality aspect of the process.
2. Do what you document—you must do your work the way you have said you will do it in the documentation.

3. Give the customer what you promised—you must have procedures for testing, inspecting, and controlling your printing processes.

Contents of ISO Standards

Clause/Title	ISO 9001	ISO 9002	ISO 9003
0 ***Introduction***	①	①	①
1 ***Scope***	①	①	①
2 ***Normative references***	①	①	①
3 ***Definitions***	①	①	①
4 ***Quality system requirements***			
4.1 Management responsibility	①	①	②
4.2 Quality system	①	①	②
4.3 Contract review	①	①	①
4.4 Design control	①	—	—
4.5 Document and data control	①	①	①
4.6 Purchasing	①	①	—
4.7 Control of customer-supplied product	①	①	①
4.8 Product identification and traceability	①	①	②
4.9 Process control	①	①	—
4.10 Inspection and testing	①	①	②
4.11 Control of inspection, measuring, and test equipment	①	①	①
4.12 Inspection and test status	①	①	①
4.13 Control of nonconforming product	①	①	②
4.14 Corrective and preventative action	①	①	②
4.15 Handling, storage, packaging, preservation, and delivery	①	①	①
4.16 Control of quality records	①	①	②
4.17 Internal quality audits	①	①	②
4.18 Training	①	①	②
4.19 Servicing	①	①	—
4.20 Statistical techniques	①	①	②

Key:
① Full requirement
② Less stringent requirement than in 9001 or 9002
— Not applicable

ISO Philosophy

- Say what you do
- Do what you say
- Document what you do
- Check the results
- Correct the difference

This is not to say that implementing ISO 9000 is easy; it isn't. However, many printing organizations are already doing many of the things necessary for certification. They must take the next step, which is to codify the tasks being performed in the form of agreed-upon standard operating procedures.

How Is ISO 9000 Implemented?

Implementing ISO 9000 is primarily a process of organizing, training, and documentation. Depending on the present level of your company's procedures and level of documentation and system's complexity, the process can take from several months to several years, with a typical registration cycle taking from twelve to eighteen months. Some printing organizations, such as in-plants for pharmaceutical companies, have been documenting their procedures and policies for years. For them, it should be very easy to get ISO 9000 certification. Others, including many commercial printers, have done a very poor job of writing down what they do. For them, it will take longer to become certified. A basic plan of attack for getting certified would consist of four phases.

Four Phases of ISO Registration

- Management commitment
- Training and organization
- Documentation
- Third-party audit

Phase I. In phase I, the company's top management becomes committed to pursuing ISO 9000. This involves training top managers as to what ISO 9000 is, how much it will cost, why it should be pursued, and what will be their role in the process. Top management's primary role is to allocate resources, set goals, review progress, and develop an awareness throughout the organization that ISO 9000 is an important pursuit as well as making it very clear that everyone will be held accountable for knowing and implementing the quality system as it affects their area.

Phase II. Phase II of the implementation is to develop a structure and a system that will support the certification process. It will be necessary to involve employees at all levels and from all functions of the organization in the process. Various leaders, such as department heads, must be trained in the methods of documentation and standard operating procedure (SOP) development. They, too, must be made aware why the process is important and how the benefits will make the organization function better. At this point, it is a good idea to train cross-functional teams that can audit each department. This will greatly assist you in determining what needs to be accomplished for each of the ISO criteria.

Phase III. In phase III of the implementation, the actual SOP development begins. The outcome of this process is typically a four-tiered system of documentation consisting of a quality manual, which is a set of company policies concerning quality. The second tier is process specifications that document what should be done and who will do it. The third is the development of work instructions. The fourth tier consists of quality records indicating that a procedure has been carried out and includes a calibration log and completed purchase orders as examples of this process. This phase is by far the most time-consuming and arduous for most companies, especially if documentation is minimal to begin with or functional areas such as prepress and press don't have a strong tradition of communicating very well with one another.

The following diagram illustrates the four-tiered quality system documentation typically associated with ISO 9000:

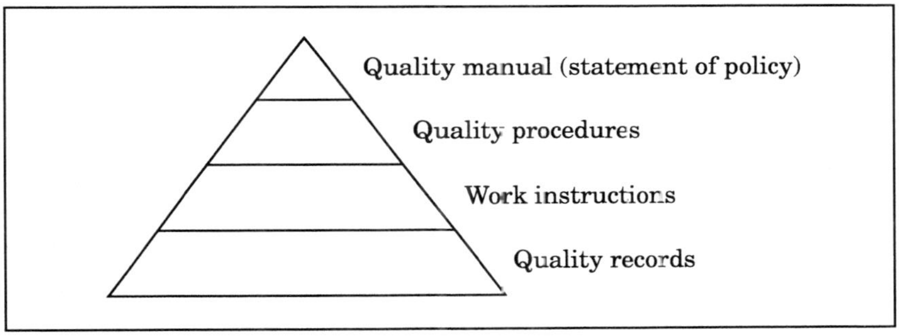

Phase IV. The final stage is phase IV, in which an outside auditor (registrar) is brought in for an ISO 9000 quality audit. It is beneficial to have a pre-audit done by the third-party auditor. The auditor will evaluate your ISO system pointing out its weaknesses and strengths. They will not assist you with suggestions on how to fix deficient areas. During the official audit, the auditor will look at your documentation developed in the previous phases and then go out into the plant and observe what the people are actually doing. If the procedure says that each plate will have a UGRA scale burned on it and be checked for proper exposure, the auditor will simply check if the platemaker is doing this and if the records are there to support that it is being done on the prescribed basis. If all procedures are found to be followed, the printer will be issued a certificate that the company is in conformance with ISO 9000. This assumes that the system itself meets all of the required elements of the standard.

The Quality Manual

A quality manual should be organized as the backbone of the quality system. It should be written in such a manner that it will not have to be continually revised. Quality manuals rarely exceed thirty pages in length. When organizing your quality manual, it is best to follow the sequence as set forth by the ISO 9000 standard that your company chooses to follow. The table of contents of a quality manual might look like the example shown at the top of the next page.

The following are two examples of how printers might write up sections in their quality manual.

Quality Manual Writing Procedure

1 Scope
This *Quality Manual* describes the quality assurance system at Casey Press as derived from the ISO 9002 and 9004 standards.

4.5 Document and Data Control
Controlled copies of the *Quality Manual* and *Instruction Handbook* are printed with the company logo and dated. These manuals and handbooks are provided to all managers and supervisors. Each copy is clearly numbered. These numbers are recorded and kept by the quality control manager, who also controls the documents in terms of revisions and appropriate approvals.

Casey Press Quality Manual
Table of Contents

ISO 9002 Clauses	Topic	Page
0	History	1
1	Scope	2
2	References	3
3	Glossary of Key Terms	4
4.1	Management Responsibility	6
4.3	Contract Review	7
4.5	Document and Data Control	8
4.6	Purchasing	9
4.7	Control of Customer-Supplied Products	11
4.8	Product Identification and Traceability	12
4.9	Process Control	13
4.10	Inspection and Testing	15
4.11	Control of Inspection, Measuring and Test Equipment	16
4.12	Inspection and Test Status	18
4.13	Control of Nonconforming Product	20
4.14	Corrective and Preventative Action	22
4.15	Handling, Storage, Packaging, Preservation, and Delivery	23
4.16	Control of Quality Records	25
4.17	Internal Quality Audits	26
4.18	Training	27
4.19	Servicing	28
4.20	Statistical Techniques	30

Standard Operating Procedures

When writing standard operating procedures for ISO 9000, it is very important that certain items be given consideration. In particular, it is critical to identify who will perform the procedure, who is responsible for enforcing the procedure, what the actual procedure is, how people get trained in the procedure, how frequently the procedure is to be performed, and what types of records are associated with the procedure. A typical ISO 9000 procedure format would have the components shown in the table at the top of page 302.

These components form the basis for the standard operating procedure that the auditor observes and the operator follows. The ultimate goal is to have an SOP that, when followed, results in a product that will be made properly the first time, every time. An example of an SOP for plate checking is found on the next page.

Casey Press Standard Operating Procedure

- **Name of the procedure:** Plate checking
- **Purpose of the procedure:** To verify plate accuracy before going to press
- **Scope of the procedure:** The platemaking department
- **References:** A list of documents or manuals that support the process—i.e., checklists, instruction guides, log books, etc.
- **Definitions:** Terms used in the procedure and their meanings
- **Responsibility:** The platemaker
- **Procedure:** What, when, and how the procedure is to be performed—i.e., after processing, the platemaker checks the plate, using a collimated 10× magnifying glass, for visual defects
- **Records:** How are records to be kept and who should keep them, for how long—i.e., a checklist is filled out by the platemaker identifying defects found and corrective actions
- **Attachments:** Documents that support the process—i.e., a flow chart

Types and Purposes of Audits

It is worth taking a few moments to discuss the subject of auditing as it relates to ISO 9000, as this is a key process in the standard. ISO 9000 mandates that the company audit its own processes internally as well as being audited regularly by an outside firm. There are generally three different types of audits as it pertains to who is doing the audit and two different types of audits as it pertains to what is being audited. These are:

- ***Who is doing the audit:***
  1. First party—a company auditing itself; internal audit
  2. Second party—a company auditing its supplier or audited by its customer
  3. Third party—a company audited by an independent outside auditing body
- ***What will be audited:***
  1. Adequacy assessment—the system meets the criteria of the standard
  2. Verification assessment—the system is being implemented as it is spelled out

Under ISO 9000, companies will audit themselves internally on a periodic basis. This is known as first-party audits. This audit consists of a review of the procedures and how they are being implemented; a department or person who is independent of the department or person being audited performs the review. Under ISO, the company must schedule, perform, follow up on, and keep records of these audits. This can be a tremendous amount of work but is also probably the most effective process improvement tool in the ISO tool box because it forces the company to constantly evaluate itself and how it does business. The use of cross-functional teams works very well in this process. (Refer to the section on team training in Chapter 9.)

Also under ISO is the requirement that to become certified, a company is audited by an independent firm that specializes in ISO registration processes. These companies will typically first conduct an adequacy assessment in which they ask you for copies of your quality system documentation. These materials are reviewed and compared against the requirements of the standard. If your system isn't found to comply, on paper, with ISO 9000, it would be impossible to actually go on to the next phase of auditing in which implementation of your system is evaluated. If the adequacy audit is passed, the assessor will conduct an audit of how well you actually implement what you say you do. This is typically done on-site by a team of auditors over a period of several days. Again, the third-party auditor will first audit what you say you do against what the standard requires and then audits what you actually do against what you say you do.

Suggested ISO Certification Audit Process

- First-party internal audit
- Third-party paper pre-audit
- Third-party on-site pre-audit
- Certification audit
- Ongoing first-party internal audit
- Ongoing third-party internal audit

First-party audit preparation. After choosing a cross-functional team and doing the appropriate team building exercises as outlined in Chapter 9 ("The Team Approach"), it will be necessary to develop questions for each clause. Remember that auditors can not audit their own departments. As an example, if the team was reviewing the criteria for Internal Quality Audits (4.17 of ISO 9002), some appropriate areas for investigation would include, whether:

1. Documented procedures cover how to carry out, document, and follow up on planned, comprehensive, internal audits of both the quality system procedures and quality results.
2. An adequate number of trained, independent people are available to implement the auditing procedures.
3. Audits are conducted at defined intervals.
4. Records indicate that the audits have been useful in driving improvement.
5. Records indicate that improvements recommended are made in a timely manner.
6. Records indicate that improvements are effective.
7. Records indicate that senior management reviews the results of internal audits on a regular basis.

The internal auditing team should grade the investigation as follows:

0 Does not apply at all
1 There has been discussion about the need for this clause
2 Rough drafts or outdated versions exist
3 Documentation exists, but it is superficial, seen as insufficient, or seen as not useful
4 Documentation exists and is generally useful, but it is not always kept up to date
5 Relevant documentation is available, complete, useful, and up to date.

What Are the Benefits?

There are many benefits to using the ISO 9000 process of documentation and auditing. While your company might not sell to Europe or be required to become certified, there are still ample reasons to pursue this quality standard. A few of these include:

- ***Jump starting and managing the quality improvement process.*** Many printers have only a vague idea of what total quality is and how they should pursue it. ISO 9000, while never intended to be a continuous improvement process or a total quality model, such as the Baldrige Award, forces the company to take a disciplined look at its present practices and to begin the first phase of implementing total quality—the development of SOPs. This helps the company get its processes under control.
- ***Breaking down organizational boundaries.*** The ISO 9000 process forces printers to work across functions and levels. Developing the necessary documentation will

necessitate working with various departments such as bindery and shipping in order that truly workable specifications are developed. It would be impossible to implement ISO 9000 without a total, company-wide, team effort.

- ***Augmenting the training process.*** By developing work documentation, the printer also develops a database of the necessary skills and procedures for getting the job done properly. This information can be used for craft training when hiring or promoting employees to new positions.
- ***Marketing rewards.*** Many customers are beginning to look at the quality processes employed by their printers. They don't want to inspect your work; instead, they expect you to deliver a defect-free product. ISO 9000 can provide assurance that your system is capable of producing consistent results. This can be a useful selling point and a way of differentiating your company from the pack. Additionally, while ISO 9000 might not produce perfect results, it is also not "smoke and mirrors." If you say you do something under ISO 9000, you better do it or you will either never gain certification or have your certification revoked.
- ***Costs.*** The British Standards Institution reports that operating costs, on average, decrease by approximately 10% for companies that implement ISO 9000. Savings come in the form of reduced defects, less redundant work procedures, and lower costs of inspection. These are, of course, reductions in failure costs as a payback on your prevention and appraisal dollars.
- ***Supplier evaluation.*** Many suppliers such as ink or paper manufactures have either implemented ISO 9000 or are looking at the process. The wise printer would be well served by asking whether the supplier is into ISO 9000, and if not, asking "why not?" Use the criteria in ISO 9004 for deciding whether your supplier is capable of meeting your needs from the quality standpoint. It is perfectly legitimate to ask your ISO 9000 supplier to provide you with a copy of its quality manual as part of your supplier survey process. While this will not be detailed, there will nevertheless be much useful information in the manual for you to evaluate your supplier on.
- ***Other benefits.*** Consistency of operations, potential for improved quality, improved customer perceptions of quality, potential new customers, lower costs leading to higher profits, improved market share, less rework, lower waste and spoilage, and reduced inspection costs.

The sample survey questionnaire on page 307 shows ISO-based questions you might ask your supplier.

Cost of Implementation

An average size printing company should allocate 1% of its sales dollar in preparation for an 18-month to 24-month plan to become ISO-certified. However, approximately only 27% of this 1% allocation will be directly out-of-pocket expenses; 73% will be the cost of employee involvement in time.

The costs of implementing ISO 9000 have been estimated to run as low as $10,000 to $15,000 actual out-of-pocket expenses, although a more accurate figure that takes into account man-hours to document and training time would be in the $200,000 range. ISO 9000 can be a costly process when you start adding up the internal work that must be done to build and maintain the system (SOP development, administration, record keeping, training, internal auditing). However, if you look at the costs and the benefits, it is likely that you will conclude that ISO 9000 is a good chance to start your quality improvement process and is well worth the money.

ISO Cost Estimates

Category	Percent of Cost
Registrar	9
Outside consultant	18
Internal	73

Summary

Keep in mind that ISO 9000 is not the "ultimate quality process" as much as the "minimal requirements of a quality system." What this means is that ISO is a good way to get started in your quality improvement journey. Even for a company that is well into its quality journey, ISO might be a good way to institutionalize your improved processes so that you minimize the risk of going back to the old ways of doing things. Basically, ISO 9000 allows you to get your house in order. This chapter has discussed ISO 9000 from the general standpoint of how it might be useful to a printing company, keeping in mind that the standard itself is revised every five years. The company interested in pursuing ISO 9000 should contact the Registrar Accreditation Board (RAB) through ASQC. The RAB and ASQC can provide you with information about the standard and its application as well as pertinent books and a copy of the standard itself.

Supplier Questionnaire

Company Name: ______________________________ Date: ________________________

Person Responding: __________________________ Title: ________________________

Please respond to the following questions regarding your system for assuring quality:

	Yes	No	In Progress	Don't Know
1. We have a quality policy in place.	☐	☐	☐	☐
2. We have stated quality objectives.	☐	☐	☐	☐
3. We have an established quality system.	☐	☐	☐	☐
4. Our quality system is fully documented.	☐	☐	☐	☐
5. Our operational procedures are documented.	☐	☐	☐	☐
6. We have a quality manual.	☐	☐	☐	☐
7. We actively plan for quality improvement.	☐	☐	☐	☐
8. We maintain quality records.	☐	☐	☐	☐
9. We measure quality costs.	☐	☐	☐	☐
10. We have complete product specifications.	☐	☐	☐	☐
11. We pursue long-term supplier relations.	☐	☐	☐	☐
12. We measure/test incoming products.	☐	☐	☐	☐
13. We use SPC for controlling quality.	☐	☐	☐	☐
14. We have nonconformance policies.	☐	☐	☐	☐
15. Our personnel are trained in SPC.	☐	☐	☐	☐
16. We have ISO 9000 certification.	☐	☐	☐	☐

Please describe in general terms your approach to quality assurance and improvement: ____

____

____

____

____

____

____

____

Please provide documentation, where possible, related to the above questions.

Glossary

acceptance sampling — Inspection of a sample from a lot to decide whether to accept or not accept that lot. The sampling can be an attribute (go/no-go) or a variable.

acceptance sampling plan — A specific plan that indicates the sampling sizes and acceptance or non-acceptance criteria to be used.

affinity diagram — A brainstorming organizational technique that groups ideas into those items that are related.

analysis of variance — A statistical technique for analyzing experimental data.

AOQ — Average outgoing quality; for a given sampling plan and incoming product quality, it is the average quality level of outgoing product.

AOQL — Average outgoing quality limit; for a given sampling plan, it is the worst possible outgoing quality over all possible levels of incoming quality.

AQL — Acceptable quality level; the fraction defective that can be tolerated without serious effect upon further processing operations or customer reaction. It is commonly designated by the symbol p_1.

ASQC — American Society for Quality Control; a professional association that develops, promotes, and applies quality-related information and technology.

assignable cause A cause of quality variation that is identifiable and that may be eliminated or controlled economically (also known as a "special cause.")

attribute data Qualitative data that can be counted for recording and analysis. Attribute data are usually gathered in the form of nonconforming units or of nonconformities; they are analyzed by p, np, c, and u control charts.

attribute measurements One in which a quality characteristic is present or absent, or falls into a limited number of discrete categories.

average The sum of values divided by the number (sample size) of values, and designated by x-bar. (See also "mean.")

average chart A control chart in which the subgroup average, x-bar, is used to evaluate the average output of the process.

benchmarking An improvement process in which performance of a company is measured and compared to the best-in-class companies.

big Q, little q Contrasts the difference between managing for quality in all business processes and products (big Q) and managing for quality in a limited capacity, i.e., manufacturing (little q).

bimodal distribution Used in analyzing a histogram to determine whether there are two influences upon the data.

binomial distribution Describes a population or process where each item in the population has exactly one of two values (e.g., go/no-go; on/off; good/bad).

blocking The systematic isolation of differences between various experimental conditions. A technique used in design of experiments (DOE).

brainstorming A technique used by teams to generate ideas on a particular subject.

c chart The chart that is used to monitor the number of defects in a given sample size.

calibration Comparing a measurement instrument of unverified accuracy to a measuring instrument of known accuracy to detect variation from the required performance specification.

capability When the process average is ±3 standard deviations, the spread of the distribution of individuals is contained within the specification tolerance (variable data), or when at least 99.73% are within specification (attribute data), the process is capable. This can only be determined after the process is in statistical control.

cause-and-effect diagram Illustrates the main causes and subcauses that lead to an effect. Also referred to as the "Ishikawa diagram" or "fishbone diagram."

center line The line on a control chart that represents the average or mean value of the items being plotted.

chain reaction Dr. Deming observed that by improving and focusing on quality you would be able to reduce waste, which would in turn reduce costs, improve pricing flexibility, improve turnaround time, and therefore improve customer satisfaction.

chance causes Causes of quality variation that can seldom be identified and that are never economical to control or eliminate. Individually their effect is negligible; collectively they make up the inherent variability in a constant cause system (also known as "common causes").

characteristic A distinguishing feature of a process or its output on which variable data can be collected.

check sheet A simple data-recording device, custom-designed by the user, that allows for results to be readily interpreted.

checklist A tool to ensure that all important steps or actions in an operation have been taken correctly.

common cause A source of variation that affects all values of the process output being studied; in control chart analysis, it appears as part of the random process variation.

conformance A positive indication or judgment that a product or service has met the requirements or critical specification.

consecutive units Units of output produced in succession; a basis for selecting subgroup samples.

contingency method A brainstorming approach that requires participants to look at the negative side of the situation: what can be done to defeat a process. This then can be turned around to improve the process.

continuous improvement The ongoing improvement of products, services, or processes through incremental and breakthrough improvements.

control chart A graphic representation of a characteristic of a process, showing plotted values of some statistic gathered from that characteristic, a center line, and two control limits. It has two basic uses: as a judgment to determine if a process has been operating in statistical control, and as an operation to aid in maintaining statistical control.

control limit Lines on a control chart used as a basis for judging the significance of the variation from subgroup to subgroup. Variation beyond a control limit is evidence that special causes are affecting the process. Control limits are calculated from process data and are not to be confused with engineering specifications.

corrective action Implementation of solutions resulting in the reduction or elimination of an identified problem.

cost of quality The costs associated with and preventing poor quality.

count chart A control chart for evaluating the stability of a process in terms of the count of events of a given classification occurring in a sample.

Cp Process capability ratio; is defined as the ratio of the tolerance spread to the total spread of the process.

Cpk Process capability index; also shows how well the process is centered on its nominal dimension.

customer See "external customer" and "internal customer."

customer-supplier partnership A long-term relationship between a buyer and a supplier characterized by teamwork and mutual confidence.

decision matrix A matrix used by teams to evaluate problems or possible solutions.

defect A failure to meet a specified quality standard.

defective An item that contains one or more defects.

dependent variable Used in correlation diagrams to indicate the variable that is affected by the independent variable and is plotted on the y-axis of the graph.

detection A past-oriented strategy that attempts to identify unacceptable output after it has been produced and to separate it from the good output. (See also "prevention.")

distribution A description of a process's average output and the variation of output from that average.

DOE Design of experiments; the planning, conducting, analyzing, and interpreting of controlled tests to evaluate inputs and outputs of a process.

85:15 rule Dr. Deming defined this rule by observing that 85% of quality is a function of management with only 15% being operator controllable.

80-20 rule Eighty-twenty; a term first defined by J. M. Juran that suggests that 80% of the effects come from 20% of the possible causes (also known as the "Pareto rule").

empowerment A process whereby employees regularly participate in making decisions on how their work areas operate.

EQC Executive quality council; a permanent team that coordinates team efforts, authorizes projects, gives financial support, and has authority for final decisions as to what will be implemented.

EVOP Evolutionary operation; moves all variables simultaneously and looks at the interactions between them.

external customer	A person or organization that receives a product or service but is not part of the organization supplying it.
fishbone diagram	See "cause-and-effect diagram."
flowchart	A graphical description of the activities, decisions, inputs, and outputs involved in a given process.
histogram	A graph that gives a graphic summary of the variation in a set of data. The pictorial nature of the histogram lets people see patterns that are difficult to see in a simple table of numbers. (See "distribution.")
independent variable	Used in correlation diagrams to indicate the variable that is causing the problem and is plotted on the x-axis of the graph.
individual	A single unit or single measurement of a characteristic.
inspection	Measuring, examining, testing, or gauging one or more characteristics of a product or service and comparing the results with specified requirements to determine whether conformity is achieved.
internal customer	The recipient (person or department) of another person's or department's output (product or service) within an organization.
Ishikawa diagram	See "cause-and-effect diagram."
ISO 9000 series standards	A set of five individual but related international standards on quality management and quality assurance developed to help companies effectively document the quality system elements to be implemented to maintain an efficient quality system.
JIT manufacturing	Just-in-time manufacturing; a planning system for a manufacturing process in which there is little or no manufacturing material inventory on hand at the manufacturing site and little or no incoming inspection.
kaizen	A Japanese term that means gradual unending improvement by doing little things better and setting and achieving increasingly higher standards.

kurtosis Abnormal concentration of data in the shoulders or extended tail areas of the distribution. Values greater than zero indicate longer tails (platykurtosis) while values less than zero indicate fat shoulders and less tails (leptokurtosis).

LCL Lower control limit.

Malcolm Baldrige National Quality Award An award established by Congress in 1987 to raise awareness of quality management and to recognize U.S. companies that have successfully implemented quality management systems.

mean The average of values in a group of measurements.

median The middle value in a group of measurements, when arranged from lowest to highest; if the number of values is even, then by convention the average of the middle two values is used as the median.

milieu Environmental concerns that include the physical and mental situation. This is part of the cause-and-effect analysis process.

MIL-STD-105E A military standard that describes the sampling procedures and tables for inspection by attributes.

mode That value that occurs most frequently in a series of numbers.

n The number of observations or items (i.e., sample size).

NIH Not invented here; the implementer had little or no input into the process and therefore does not feel a part of what is to be implemented.

nominal dimension The specification that is aimed for.

nonconforming units Units that do not conform to a specification or other inspection standard; sometimes called "discrepant" or "defective" units.

nonconformities Specific occurrences of a condition that does not conform to specifications or other inspection standards; sometimes called "discrepancies" or "defects."

normal distribution A continuous, symmetrical, bell-shaped frequency distribution for variable data that underlies the control charts for variables and the process capability study.

OC curve Operating characteristic curve; a curve that graphically portrays the probability of acceptance of a lot according to the fraction defective in the lot, when sampled by a given sampling plan.

operational definition A means of clearly communicating quality expectations and performance; it consists of (1) a criterion to be applied to an object or to a group, (2) a test of the object or of the group, and (3) a decision as to whether the object or the group met the criterion.

OVAT One variable at a time.

p chart Percent chart; also known as a "proportion chart," it is a control chart for monitoring the total number of defectives as a proportion (or percent) in a sample.

Pareto chart A simple tool for problem solving that involves ranking all potential problem areas or sources of variation according to their contribution of cost or to total variation.

PDCA cycle Plan-Do-Check-Act cycle; also referred to as the "Shewhart cycle," it is a four-step process for quality improvement. A ***plan*** to effect improvement is developed in the first step, and it is carried out ***(do)*** on a small scale in the second step. In the ***check*** step, the effects are observed, and in the last step ***(act),*** the results are studied to determine what should be done.

prevention A future-oriented strategy that improves productivity and quality by directing analysis and action toward correcting the process itself. (See also "detection.")

probability Statistical determination of the likelihood of a random event occurring. Used in a process capability study to predict what percentage of a product is likely to be in/out of specification.

problem solving The process of moving from symptoms to causes (special or common) to actions that improve performance.

process The combination of people, equipment, materials, methods, and environment that produces output—a given product or service.

process average The central tendency of the distribution of measured values of a particular process characteristic.

process capability The determination if a process in capable of producing all of its product within designated specifications. Process capability assumes that the process is first in a state of statistical control.

process control See "statistical process control."

process mapping Same meaning as flow charting. See "flowchart."

process spread The extent to which the distributions of individual values of the process characteristic vary.

quality The characteristics of a product or service that bear on its ability to satisfy stated or implied needs or a product or service free of deficiencies.

quality assurance All the planning and systematic actions that can be demonstrated which will provide confidence that a product or service will fulfill the requirements for quality.

quality circles Quality improvement or self-improvement study groups composed of a small number of employees (ten or fewer) and their supervisor.

quality control The operational techniques and activities used to fulfill the requirements for quality.

quality function deployment Often referred to as listening to the voice of the customer. It is a structured method in which customer requirements are translated into appropriate technical requirements for each stage of product development and production.

R Range; the difference between the largest and the smallest number in a group of numbers or the highest and lowest values in a subgroup.

R chart Range chart; a control chart in which the subgroup range, "R," is used to evaluate the variability within a process.

RAB Registration Accreditation Board; private companies that are registered with the RAB as third-party registrars for ISO 9000.

random sampling A sampling technique in which sample units are selected in such a manner that each unit has an equal probability of being selected.

re-engineering The use of radical thinking to redesign business processes in order to achieve dramatic improvements in cost, quality, service, and speed. Examples can include the use of CD-ROM instead of traditional printing to communicate or empowering employees to take over traditional management responsibilities through self-directed work teams.

regression analysis A statistical technique for determining the best mathematical expression describing the relationship between dependent and independent variables.

run A consecutive number of points consistently increasing or decreasing, or above or below the center line of a control chart.

run chart A simple graphic representation of a characteristic or a process showing plotted values of some statistic gathered from the process (often individual values) and a center line (often the mean of the values) that can be analyzed for runs. (See also "control chart.")

sample In process control applications, a synonym for a subgroup, group of subgroups, or other measurements taken from a process.

scatter diagram A graphical technique to analyze the relationship between two variables. Two sets of data are plotted on a graph that will show possible relationships.

SCP Supplier certification process; the four stages of supplier partnerships where the supplier has the opportunity to prove that it can perform at an expected quality level.

shape A general concept for the overall pattern formed by a distribution of values.

sigma (s) The Greek letter used to designate the standard deviation of a sample.

sigma (Σ) The Greek letter used to designate the sum of a series of data.

skewness Abnormal concentration of data to one or the other side of the mean of a distribution.

SOP Standard operating procedure; a method of performing work that is consistent from person to person, day to day, and job to job.

SPC Statistical process control; the use of statistical techniques such as control charts to analyze a process or its outputs so as to take appropriate action to achieve and maintain a state of statistical control and to improve the process capability.

special cause A source of variation that is intermittent, unpredictable, unstable; sometimes called an "assignable cause."

specification The requirements to which a given product or service must conform to meet a given set of needs.

spread A general concept for the extent by which values in a distribution differ from one another. Also called "dispersion." Mathematically characterized by the range or standard deviation.

stable process A process in statistical control. Lacking special causes of variation.

standard deviation A measure of the variability of a process.

statistic A value calculated from or based upon sample data and used to make inferences about the process that produced the output from which the sample came.

subgroup One or more events or measurements used to analyze the performance of a process.

TQM	Total quality management; a management approach to long-term success through customer satisfaction. TQM is based on the participation of all members of an organization in improving processes, products, services, and the culture they work in. It is an all-inclusive term that encompasses all the theories and methods available for company success through customer satisfaction.
UCL	The upper control limit.
variable data	Quantitative data, where measurements are used for analysis.
variation	The inevitable differences among individual outputs of a process; the sources of variation can be grouped into two major classes: common causes and special causes.
vital few, useful many	A term used by J. M. Juran to describe the use of the Pareto principle; where 20% are considered the vital few, and the remaining 80% are considered the useful many. Refer to "Pareto chart."
why technique	A procedure to search deeper into a process. Asking "why" at least five to seven times to discover the root cause of a problem.
world-class quality	A term that refers to a standard of excellence and infers the best of the best.
x-bar chart	Average chart; used for plotting and evaluating subgroup averages over time.
zero defects	A performance standard developed by Philip B. Crosby which states that if people commit themselves to watching details and avoiding errors they can move closer to the goal of zero defects.

Resources

American Management Association (AMA), 9 Galen Street, Watertown, MA 02172

American National Standards Institute (ANSI), 1430 Broadway, New York, NY 10018

American Newspaper Publishers Association (ANPA), P.O. Box 17407, Dulles International Airport, Washington, DC 20041

American Society for Quality Control (ASQC), P.O. Box 3066, Milwaukee, WI 53201-3066

American Society for Testing and Materials (ASTM), 1916 Race Street, Philadelphia, PA 19103

Association for Graphic Arts Training (AGAT), c/o RIT/T&E Center, P.O. Box 9887, Rochester, NY 14623

Association for Quality and Participation (AQP), 801-B W. 8th Street, Suite 501, Cincinnati, OH 45203

The Association for Suppliers of Printing and Publishing Technologies (NPES), 1899 Preston White Drive, Reston, VA 22091

Binding Industries of America/Printing Industries of America (BIA/PIA), 200 E. Ontario Street, Chicago, IL 60611

Book Manufacturers Institute (BMI), 111 Prospect Street, Stamford, CT 06901

CEEM Information Services, 10521 Braddock Rd., Fairfax, VA

Center for Quality and Applied Statistics, Rochester Institute of Technology, P.O. Box 9887, Rochester, NY 14623-0887

Dunn Technology, Inc., 759 East Vista Way, Vista, CA 92083

Federal Quality Institute (FQI), P.O. Box 99, Washington, DC 20044-0099

Flexographic Technical Association (FTA), 900 Marconi Avenue, Ronkonkoma, NY 11779

FOGRA (German Research for Printing), Streitfeldstrasse 19, Postfach 80 04 69, D8000 Munchen, 80 Germany

Graphic Arts Publishing (GAP), 3100 Bronson Hill Road, Livonia, NY 14487

Graphic Arts Technical Foundation (GATF), 4615 Forbes Avenue, Pittsburgh, PA 15213

Graphic Communications Association (GCA), 100 Daingerfield Road, Alexandria, VA 22314-2888

Graphic Services+Seminars, P.O. Box 13723, San Luis Obispo, CA 93406

Gravure Association of America (GAA), 1200-A Scottsville Road, Rochester, NY 14624

GOAL/QPC, 13 Branch Street, Methuen, MA 01844

INFO, P.O. Box 606, Ayer, MA 01432

International Publishing Management Association (IPMA), 1205 West College Street, Liberty, MO 64068-3733

International Association of Quality Circles, P.O. Box 30635, Midwest City, OK 73140

International Association of Research Institutes for the Graphic Arts Industry (IARIGAI), 18 Ridgeway, Fetcham Park, Leatherhead, Surrey 6, KT22 9AZ, United Kingdom

International Business Forms Industries (IBFI), 2111 Wilson Blvd., Arlington,VA 22201

International Organization for Standards (ISO), 1, Rue D Varembe, ch-1211, Geneva 20, Switzerland

International Prepress Association (IPA), 7200 France Avenue South, Suite 327, Edina, MN 55435

Japan Printing Academy (JPA), Koishikawa 4-13-2, Bunkyo-ku, Tokyo, Japan

Joiner Associates Inc., 3800 Regent Street, P.O. Box 5445, Madison, WI 53705-0445

Juran Institute, 11 River Road, Box 811, Wilton, CT 06897-0811

Kaizen Institute, 701 Dragon, Austin, TX 78734

Magazine Publishers Association (MPA), 575 Lexington Avenue, New York, NY 10022

Malcolm Baldrige National Quality Award (MBNQA), United States Department of Commerce, Technology Administration, National Institute of Standards and Technology, Route 270 and Quince Orchard Road, Administration Building, Room A537, Gaithersburg, MD 20899

National Association of Printers and Lithographers (NAPL), 780 Palisade Avenue, Teaneck, NJ 07666

National Association of Printing Ink Manufacturers (NAPIM), 777 Terrace Avenue, Hasbrouck, NJ 07606

National Association of Quick Printers (NAQP), 401 North Michigan Avenue, Chicago, IL 60611-4267

National Business Forms Association (NBFA), 433 E. Monroe Avenue, Alexandria, VA 22301

National Printing Ink Research Institute (NPIRI), Lehigh University, Bethlehem, PA 18015

Naval Publications and Forms Center, 5801 Tabor Avenue, Philadelphia, PA 19120

Philip B. Crosby Associates, Inc., 3260 University Blvd., Suite 175, Winter Park, FL 32792

Printing Industries of America (PIA), 100 Daingerfield Road, Alexandria, VA 22314-2888

Quality Circle Institute, P.O. Box Q, Red Bluff, CA 96080

Quality Control Scanner. Graphic Arts Publishing Co., 3100 Bronson Hill Road, Livonia, NY 14487

Quality Digest. P.O. Box 882, Red Bluff, CA 96080

Quality Progress. American Society for Quality Control, P.O. Box 3005, Milwaukee, WI 53201-3005

Registration and Accreditation Board (RAB), American Society for Quality Control, P.O. Box 3066, Milwaukee, WI 53201-3066

Research and Engineering Council of the Graphic Arts Industry (R&E Council), P.O. Box 639, Chadds Ford, PA 19317

Research Association for the Paper and Board, Printing and Packaging Industries (PIRA), Randalls Road, Leatherhead, Surrey, KTSS 7RU, England

Screenprinting and Graphic Imaging Association, 10015 Main Street, Fairfax, VA 22031-3489

Technical and Education Center of the Graphic Arts (T&E Center), Rochester Institute of Technology, One Lomb Memorial Drive, Rochester, NY 14623

Technical Association of Pulp and Paper Institute (TAPPI), P.O. Box 105113, Atlanta, GA 30348-5113

Technical Association of the Graphic Arts (TAGA), RIT/T&E Center, One Lomb Memorial Drive, P.O. Box 9887, Rochester, NY 14623

U.S. General Accounting Office (GAO), P.O. Box 6015, Gaithersburg, MD 20877

Bibliography

American National Standard: Definitions, Symbols, Formulas, and Tables for Control Charts. Milwaukee, WI: American Society for Quality Control, 1987.

American National Standards: Quality Management and Quality Assurance Standards—Guidelines for Selection and Use. ANSI/ASQC Q9000. Milwaukee, WI: American Society for Quality Control, 1994.

American National Standards: Quality Management and Quality System Elements—Guidelines. ANSI/ASQC Q9004. Milwaukee, WI: American Society for Quality Control, 1994.

American National Standards: Quality Systems—Model for Quality Assurance in Design, Development, Production, Installation, and Servicing. ANSI/ASQC Q9001-2-3. Milwaukee, WI: American Society for Quality Control, 1994.

Artyer, Dennis R. *Quality Audits for Improved Performance.* Milwaukee, WI: ASQC Quality Press, 1989.

The ASQC Basic References in Quality Control: Statistical Techniques. Milwaukee, WI: ASQC Quality Press, 1986–89.

Barker, A. Thomas. *Quality by Experimental Design*. New York, NY: Marcel Dekker, Inc., 1985.

Blake, Robert. *Spectacular Teamwork*. New York, NY: John Wiley and Sons, 1987.

Bossert, James L. *Procurement Quality Control.* Milwaukee, WI: ASQC Quality Press, 1988.

Box, George E. P., and Norman R. Draper. *Evolutionary Operation: A Statistical Method for Process Improvement.* New York, NY: John Wiley and Sons, 1969.

Byars, Lloyd L., and Leslie W. Rue. *Human Resource Management, Fourth Edition.* Homewood, IL: Richard D. Irwin, Inc., 1994.

Byham, William C., and Jeff Cox. *Zapp! The Lightning of Empowerment.* New York, NY: Harmony Books, 1990.

Campanella, Jack. *Principles of Quality Costs.* Milwaukee, WI: ASQC Quality Press, 1990.

Clemmer, Jim. *Firing on All Cylinders.* Homewood, IL: Richard D. Irwin, Inc., 1992.

Compton, John. *Quality Concepts, Volumes I and II.* Rochester, NY: Rochester Institute of Technology, 1993.

Crosby, Philip B. *Quality Is Free: The Art of Making Quality Certain.* New York, NY: McGraw-Hill Book Company, 1979.

Deming, W. Edwards. *Out of the Crisis*. Cambridge, MA: MIT, Center for Advanced Engineering Study, 1982.

Feigenbaum, Armand V. *Total Quality Control.* New York, NY: McGraw-Hill Book Company, 1983.

Gitlow, Howard S., and Shelly J. Gitlow. *The Deming Guide to Quality and Competitive Position*. Englewood Cliffs, NJ: Prentice-Hall, Inc., 1987.

Gitlow, Howard S., et. al. *Tools and Methods for the Improvement of Quality.* Boston, MA: Irwin, 1989.

Glossary and Tables for Statistical Process Control. Milwaukee, WI: American Society for Quality Control, 1983.

Goldratt, Eliyahu M., and Jeff Cox. *The Goal.* Croton-on-Hudson, NY: North River Press, 1992.

Hale, Roger L., et. al. *Managing Supplier Quality.* Exeter, NH: Monochrome Press, 1994.

Halloran, Jack, and George L. Frunzi. *Supervision: The Art of Management.* Englewood Cliffs, NJ: Prentice-Hall, Inc., 1986.

Imai, Masaaki. *Kaizen: The Key to Japan's Competitive Success.* New York, NY: McGraw-Hill Publishing Company, 1986.

Ishikawa, Kaoru. *Guide to Quality Control.* White Plains, NY: Asian Productivity Organization, 1982.

———. *What Is Total Quality Control? The Japanese Way.* Englewood Cliffs, NJ: Prentice-Hall, Inc., 1985.

The ISO 9000 Standards: A Practical Overview. Video Conference Participant Materials, New River Media, Inc., 1994.

Juran, J. M. *Juran's Quality Control Handbook.* New York, NY: McGraw-Hill Book Company, 1988.

Juran, J. M., and Frank M. Gryna. *Quality Planning and Analysis.* New York, NY: McGraw-Hill Publishing Company, 1986.

Krishnamourthi, K. *Quality Control for Operators and Foremen.* Milwaukee, WI: ASQC Quality Press, 1989.

Kume, Hitoshi. *Statistical Methods for Quality Improvement.* Tokyo: Association for Overseas Technical Scholarship, 1988.

Lerner, Joel J., and H. A. Baker. *Schaum's Outline of Theory and Problems of Introduction to Business.* New York, NY: McGraw-Hill Book Co., 1976.

Maass, Richard A., John O. Brown, and James L. Bossert. *Supplier Certification, A Continuous Improvement Strategy.* Milwaukee, WI: ASQC Press, 1990.

Malcolm Baldrige National Quality Award. Gaithersburg, MD: United States Department of Commerce, Technology Administration, National Institute of Standards and Technology, 1995.

McCloskey, Larry A., and Dennis N. Collett. *TQM: A Basic Text.* Methuen, MA: GOAL/QPC, 1993.

The Memory Jogger. Methuen, MA: GOAL/QPC, 1988.

Military Standard Sampling Procedures and Tables for Inspection by Attributes. MIL-STD-105E. Washington, DC: Department of Defense, 1989.

Military Standard Sampling Procedures and Tables for Inspection by Variables for Percent Defective. MIL-STD-414. Washington, DC: Department of Defense, 1957.

Mohr, William. *Quality Circles*. Menlo Park, CA: Addison-Wesley Publishing Co., 1983.

O'Conner, Michael P., and Becky Erickson. *The MNM Team Building Process for Printers.* Portland, ME: Old Stone Publishing, 1991.

Patchin, Robert. *The Management and Maintenance of Quality Circles*. Homewood, IL: Dow Jones-Erwin, 1983.

Paton, Scott. "Joseph M. Juran, Quality Legend." *Quality Digest.* Red Bluff, CA: January, February, March, April 1992.

Peach, Robert W. *The ISO 9000 Handbook.* Fairfax, VA: CEEM Information Services, 1992.

Perry, William E. *Malcolm Baldrige National Quality Award Newsletter*. Orlando, FL: Quality Assurance Institute, 1991.

Quality Control Manual. Ronkonkoma, NY: Foundation of the Flexographic Technical Association, Inc., 1990.

The Quality Review. The Prophets of Quality. Milwaukee, WI: ASQC Quality Press, Winter 1988.

Ross, Phillip J. *Taguchi Techniques for Quality Engineering: Loss Function, Orthogonal Experiments, Parameter and Tolerance Design.* New York, NY: McGraw-Hill Book Company, 1988.

Scherkenbach, William W. *The Deming Route to Quality and Productivity: Roadmaps and Roadblocks.* Milwaukee, WI: ASQC Quality Press, 1988.

Scholtes, Peter R. *The Team Handbook.* P.O. Box 5445, Madison, WI: Joiner Associates Inc., 1990.

Schrock, Edward M., and Henry L. Lefevre. *The Good and the Bad News about Quality.* New York, NY: Marcel Dekker, Inc., 1988.

Shewhart, A. Walter. *Economic Control of Quality Manufactured Product.* New York, NY: Van Nostrand Press Company, Inc., 1931 (Republished, Milwaukee, WI: ASQC Quality Press, 1980).

Shewhart, A. Walter. *Statistical Method from the Viewpoint of Quality Control.* Washington, DC: The Graduate School of the Department of Agriculture, 1939.

Shores, A. Richard. *Survival of the Fittest: Total Quality Control and Management Evolution.* Milwaukee, WI: American Society for Quality Control, 1988.

Sloan, David, and Scott Weiss. *Supplier Improvement Process Handbook.* Milwaukee, WI: American Society for Quality Control, 1987.

Southworth, Miles, and Donna Southworth. *Quality and Productivity in the Graphic Arts.* Livonia, NY: Graphic Arts Publishing, 1989.

Talley, Dorsey. *Total Quality Management.* Milwaukee, WI: ASQC Quality Press, 1991.

Taylor, Frederick. *Shop Management.* New York, NY: Harper, 1919.

Townsend, Patrick. *Commit to Quality.* New York, NY: John Wiley and Sons, Inc., 1986.

Wellins, Richard S., William C. Byham, and Jeanne M. Wilson. *Empowered Teams.* San Francisco, CA: Jossey-Bass Inc., 1991.

Index

About the Authors

Herschel L. Apfelberg is a professor of Graphic Communication at California Polytechnic State University. His specialties include the areas of human resources and web technology. He is also president of Graphic Services+Seminars, a training, consulting, and publishing company for the printing industry. Apfelberg spent many years in the printing, publishing, and packaging industry as a craftsman, process engineer, administrator, and manager. He consults in printing management and technology and quality issues both nationally and internationally, and he writes a monthly column for *Print-Equip News* entitled "Education and Training." Apfelberg is a frequent speaker and is the author of several books and articles. He holds a BS degree from RIT and an MA degree from California Polytechnic State University. He has done postgraduate work at the University of California at Santa Barbara. His major emphasis of academic study has been printing technology, management, and education.

Michael J. Apfelberg is manager of quality assurance with Quebecor Printing Dallas. In this capacity, he works to implement continuous improvement and total quality management. His responsibilities include surveying customers and employees, quantifying quality costs, implementing suggestion systems, using and teaching the application of statistical process control, supplier certification, specification development, and registering ISO 9000 quality systems. Before this, Apfelberg worked as the manager of quality assurance and training for the Quebecor Printing Book Group. He has held the position of industry consultant with Graphic Services+Seminars. He holds a BA and an MBA from California Polytechnic State University. A member of the American Society for Quality Control and an ASQC Certified Quality Engineer, he has regularly written and spoken on total quality topics for the printing industry.